THE STORY OF SHIT

Midas Dekkers is a bestselling Dutch writer and biologist. His books include *Physical Exercise, The Way of All Flesh, Dearest Pet* and *The Larva*.

Nancy Forest-Flier is a Dutch-to-English translator. She was educated in the USA and now lives in the Netherlands.

MIDAS DEKKERS

The Story of Shit

Translated from the Dutch by
NANCY FOREST-FLIER

TEXT PUBLISHING MELBOURNE AUSTRALIA

textpublishing.com.au

The Text Publishing Company
Swann House
22 William Street
Melbourne Victoria 3000
Australia

Copyright © Midas Dekkers 2014

English translation copyright © Nancy Forest-Flier 2018

The moral rights of Midas Dekkers and Nancy Forest-Flier to be identified as the author and translator of this work have been asserted.

All rights reserved. Without limiting the rights under copyright above, no part of this publication shall be reproduced, stored in or introduced into a retrieval system, or transmitted in any form or by any means (electronic, mechanical, photocopying, recording or otherwise), without the prior permission of both the copyright owner and the publisher of this book.

Every attempt has been made to trace copyright. Where the attempt has been unsuccessful, the publisher would be pleased to hear from the copyright owner.

First published in Dutch as *De kleine verlossing of de lust van ontlasten* by Uitgeverij Atlas Contact, Amsterdam, 2014.
First published in English by The Text Publishing Company, 2018.

Cover design by W. H. Chong
Cover images by iStock
Page design by Jessica Horrocks
Typeset by J&M Typesetters

Printed in Australia by Griffin Press, an Accredited ISO AS/NZS 14001:2004 Environmental Management System printer.

ISBN: 9781925355178 (paperback)
ISBN 9781922253767 (ebook)

A catalogue record for this book is available from the National Library of Australia.

This book is printed on paper certified against the Forest Stewardship Council® Standards. Griffin Press holds FSC chain-of-custody certification SGS-COC-005088. FSC promotes environmentally responsible, socially beneficial and economically viable management of the world's forests.

This book was published with the support of the Dutch Foundation for Literature.

Nederlands letterenfonds
dutch foundation for literature

Contents

1
Simple Pleasures 1

2
Nice and Nasty 24

3
Private Delights 55

4
Shit Central 80

5
Do-It-Yourself 110

6
What a Relief 136

7
Brown Gold 165

8
From One Anus to Another 193

9
Water and Gas 217

10
Fun and Games 246

Bibliography 270
Illustration Credits 284
Index 285

1

Simple Pleasures

It doesn't take much to enjoy yourself. You, in fact, are your own pleasure machine. Crowing with delight, you put it into operation the minute you're born. There are real little feet for kicking, eyes for looking at everything in dumbfounded astonishment, and a mouth to cram the whole world into. As for your nose, there's a set of fingers thrown in to pick it with. And when your fingers start getting too fat for picking, your nostrils just grow along with them. No toy store can beat it.

When all the growing is done, it's time for a new form of physical pleasure: sex. Back in my day ('make love not war'), sex was regarded as pleasure par excellence. Full of expectation, we brought organs into play that up until then had just been hanging there. Our highest hopes were occasionally confirmed, but not without a struggle. A regular revolution had to be fought, and it raged like a whirlwind. Compared to the French and the Industrial, the Sexual Revolution was over in the twinkling of an eye. Sex today has become as commonplace as universal suffrage or electricity from a plug in the wall. Every department store now sells the kind of flimsy ladies' underwear that you used to have to pay an arm and a leg for at your local brothel. You pluck hook-ups like daisies right from the internet. Who's bothering with wife-swapping anymore?

Sex is no longer something to get worked up about. Our little flings have moved from the bedroom to the kitchen, where we get off on juicier flesh and the voluptuous curves of blushing fruit and exotic aromas until the promises of the cookbook photos are fulfilled course by course. Molars chew, saliva slobbers, papillae quiver with excitement in sensuous intimacy. Tenderloin steak, aged port and velvety ice cream grope each other in the dark oral interior until everyone collapses, appetites sated. Even outside the home, pleasure has migrated from the southern to the northern regions of the body. Whereas in the last century there were red lights pointing the way to the bordello, now they direct us to the restaurant, where the naked ladies of the boudoir have yielded to naked poultry served on a bed of exotic lettuce. If you want to be famous, you don't get a job in the movies as a sex bomb anymore. Now you become a top chef. *'La cuisine,'* says French chef Pierre Gagnaire, *'c'est l'amour, l'art et la technique.'* And if you aren't able to master the art of seductive cooking yourself, you watch others do it. Top chefs and housewives who have cooked their way up the ladder dish out their secrets in books and on TV. Bookshops used to sell sex manuals featuring positions you'd never before attempted, but now there are cookbooks with recipes you'll never cook, full of nude photos of chickens and rabbits. Looking at the pictures is about as far as you ever get. It's all 'gastro porn', according to foodie Johannes van Dam, who once had a shop full of them. The attention being paid to food is approaching the obscene. Food is the sex of the twenty-first century.

So it's all the more remarkable that a third form of physical pleasure has remained in the shadows. While adventures in the bedroom and kitchen are openly discussed in the living room, bathroom experiences go unmentioned. Defecating is reserved for private whispers (if it's talked about at all), and the results are furtively flushed away as if a crime had been committed. When was the last time you saw your uncle or your secretary empty their bowels? From how many of your best friends' buttocks have you actually witnessed a turd making its appearance? You don't see sex every day in real life either, but it's all the more common in films and advertising. Defecation today is at the same stage that sex was during the reign of Queen Victoria. It's more done than talked about. Anyone who spends fifteen minutes a day on the toilet will have led a secret life of approximately one year by the time they die, and not a word is said about it. Books about shitting are limited to the children's department of the bookshop. The only books on the subject in the medical section are concerned with the care and behaviour of the gut. And what you may find in the gift department—humour having to do with the lower body orifices—is best given away as soon as possible, preferably not to friends.

In the seventeenth and eighteenth centuries, Dutch tiles featuring shitting figures made for a light-hearted touch, even in stately canalside houses.

You don't shit with friends, you eat with them. When asked what they like to do best during their free time, modern people are unanimous in their response: 'Eat with friends.' Free time is when you have nothing to do, and the best nothing to do is eating. Preferably eating out. Eating out is the nothingest; you don't even have to cook. You go to the theatre first, or afterwards, or maybe even not at all. People don't eat before or after the theatre anymore; they eat instead of the theatre. And thanks to the open kitchen it's easy to follow the meal's progress. But no one pays any attention to the doors in the restaurant behind which all the food and all the drinks from all the menus finally end up. Those who feel the need to go there slink away from the table like a thief in the night and then hurry back, for, while a meal that you could have polished off in ten minutes can easily be stretched out to last a whole evening, people look at you suspiciously if you've spent more than five minutes in the loo. Yet a good shit is in no way inferior to a pleasant little dinner party. It involves the same sense of satisfaction. Although defecating, unlike love, has never been the subject of poetry, although no newspaper has a regular shitting column next to the cooking column (what should we eat today that we can relish the memory of the day after tomorrow?), it's still one of life's most elementary pleasures. You don't need to take any courses—there are no diplomas for this particular skill—yet people who can't make a simple clay ashtray can produce turds to die for. If eating is simply a matter of breaking things down, shitting is a matter of putting them back together; after destruction comes creation. Many people reappear from the toilet with a barely concealed smile. What you'd really like to do is show your artwork off to everyone, but that's a no-no for grownups. Taking pleasure in both ends of your digestive tract is a privilege reserved for the very youngest. As a toddler you're taught that anything having to do with poo is dirty; the entire southern end of your body is declared a no-go zone. Part of that region is given unrestricted access during puberty, but only the northern end of the intestines is reserved for metabolic pleasure. So there's always something childish attached to enjoying your own bowel movements.

What gives eating its exalted status? What advantage does a chop or a bunch of escarole have over a load of faeces? Food looks more appetising than shit, of course, although some sausages have a suspicious shape, and many a casserole doesn't look much different coming out from going in. It's mainly a question of smell and taste. They either draw you in or put you off. But not every stench or nasty flavour is an impregnable barrier. No matter how bitter Belgian endive may taste, no matter how much sulphide gas is produced by Brussels sprouts, we eat them anyway. It's a question of upbringing and getting older. Children are simply too young for endive and Brussels sprouts. Their intestines aren't up to it and their tongues can't stand it, and rightly so. A good glass of cognac and a lovely cigar are also beyond a toddler's ability to appreciate. Even beer is enough to make a healthy child gag. At the insistence of parents (endive, Brussels sprouts) and bad friends (cognac, cigars, beer) people learn to ignore the warning signals of their noses and tongues. But there are also tastes that even children are cured of. A newborn baby doesn't see anything wrong with poo. It'll grope around in its nappy searching for nice sweet titbits. All of us have grown up big and strong with a bit of sweet poo in our diet. It takes years of upbringing to keep a child away from the snacks produced by its own body. Learning to turn up your nose at shit is just as acquired as learning to like endive.

What you can learn you can also unlearn; things that are relinquished can also be acquired. It's called culture. Finding food delicious and poo filthy is as accidental as finding poo delicious and food filthy. The Spanish surrealist Luis Buñuel understood this. In his film *The Phantom of Liberty*, five dinner guests take their places on toilets instead of chairs. To remove any doubt, the gentlemen first take down their trousers and the ladies pull up their skirts. As they engage in dinnertime conversation, they make unashamed use of their extraordinary seats. Toilet paper is passed from guest to guest with a flourish. But when the child at the table happens to mention eating, she's told that one doesn't talk about such things. When one of the guests gets hungry, he leaves the table discreetly and asks one of the staff for directions. 'Last door on the right,' he's told. There he finds a cubicle the size of a toilet and closes the door carefully behind him. Then out of a dumbwaiter appear a plate of food, a chunk of baguette and wine, which he greedily wolfs down. There's a knock at the door to which he responds with an agitated '*occupée*', and this is answered by an embarrassed '*pardon!*' The scene is world

Filmmaker Luis Buñuel turned the world on its head. Instead of a dinner party he filmed a shitting party, where eating, not defecating, was nasty.

famous, and deservedly so. Yet Buñuel had been preceded eleven years earlier by Gerard Reve in *On My Way to the End* (1963):

> Restaurants in general make me miserable because my convictions demand that an individual partake of his food in secret, alone, preferably behind a burlap curtain, and that this food consist of the very simplest sort, with lots of raw carrot, boiled horse heart and raw swede, all of it consumed from waxed wrapping paper with a bottom layer of newspaper.

He then adds, 'Eating in the presence of dozens of other people, and strangers to boot, is something I find much more lewd than engaging in the act of sex in their presence.' When it comes to defecation, however, Reve is extremely frank. In his letters to Teigetje and Woelrat, Reve's life companions, he feels the need to keep them constantly informed of the status of his latest bowel movement. His partners manage to take this for what it's worth:

> That last item is very important: the time of day, the shape, the colour, and the smell of his turd are decisive for the success, or failure, of the coming day. In addition, it makes a great deal of difference whether he produces the whole pellet in the morning all at once or distributes it in dribs and drabs throughout the entire day.

But nothing is better at demonstrating the pleasure Reve derived from shitting than his fairy tale about Quackie the Duck, from the LP *Laying It On Thicker* (1969):

> Quackie the Duck was not at all happy that his house was clean again. So Quackie went into his little kitchen, and there he crapped a great big piece of shit out of his bottom and put the shit in an aluminium saucepan. The turd that Quackie the Duck had crapped out of his arse must have been a winner, because the shit boiled up higher and higher till it boiled out of all the windows in Quackie's little house! Now finally he was happy. He stretched himself out in the middle of the hall and lay down in his own gurgling and foaming shit. Soon he fell into a deep and satisfied sleep. And if no one had turned off the flame under the pan, it would be boiling still.

So when it comes to pleasure, pooing is right up there with eating. And then there's sex. As they say in southern Germany, '*guat geschissen ist halbat gevöglt*'—'a good shit is half a shag'. By insisting that sex and shitting be done in isolation, society creates a climate of furtiveness that appeals to many. Even solo sex is generally regarded as a primal need, just like defecation—something that can be indulged in as long as you keep it to yourself. Both of them sweeten life in a natural, casual sort of way, without all the fuss and bother attached to planning a dinner party or conducting a romantic affair. While these little pleasures may not be everyone's idea of an exalted experience, a life full of secret amusements can actually be lots of fun. 'A nice big bowel movement is the common man's orgasm,' according to the Austrian magazine *Der Wiener*. But a few uncommon men have also come to appreciate its delights: 'Nothing surpasses the pleasure of a good solid crap,' Balzac once opined.

As easy as physical pleasure can be, for many people it's problematic. While babies take an instinctual delight in their bodies, adults act as if they've misplaced the instruction booklet. They just can't seem to get the hang of it. Even after years of anatomical study, physiological testing and a professorship in internal medicine, it's still difficult to feel physically at ease. Intestines and kidneys are not part of one's self-identity.

We are strangers in our own bodies. We have no idea what's going on inside us, only what goes in and what comes out. We don't inhabit our bodies, we just serve as customs agents. We don't need to know what's taking place on the domestic front as long as we keep an eye on the incoming and outgoing goods. Once you're inside the country the customs officials are no longer interested in you, and once you've left they're even less interested. Only the border itself is closely guarded. All imported items must pass through sensitive scanners. The eyes inspect potential immigrants from a distance, even before they arrive. Mounted directly above the mouth is the nose, which extends to the throat by way of the nasal passage and makes sure the taste buds have done their work properly. Even the ears join in. They listen to make sure the food has the right crunch in addition to tasting good. Only when the scary bits of the outside world have a fully stamped passport are they permitted to enter the country's interior.

Everyone guards their throat like a virgin guards her vagina. Whether it's dubious food or a dubious guy, it's all about preserving the body's integrity. Forcing something unpleasant on someone bears a suspiciously strong resemblance to rape. The most important difference is that there's a set of teeth behind the lips of the mouth which victims of sexual assault might have found helpful—behind their other set of lips, of course.

All technical accuracy aside, getting your food cleared through customs before it enters the intestines is anything but disagreeable. Rarely have the beneficial and the pleasant been so intimately combined. As far as the senses are concerned, there's only one criterion: that it taste good. If it does, then you have everything the world has to offer from which to make your choice. The sensation we call 'delicious' is an ample reward for all the trouble involved in foraging and meal preparation. Occasionally something slips in that isn't good for you, and occasionally you refuse something disgusting that's supposed to be good for you, but all in all it's a simple system that works for a lifetime. Thanks to a bit of education, and with the support of the pure food and drug authorities, there are surprisingly few cases of food poisoning, and most people think back on their meals with pleasure. If anything goes wrong it usually isn't a question of quality but of quantity. Your uvula is not fitted with a pair of scales. Good food is something many people can't get enough of.

The epicure may rub his belly with contentment, but there's nothing more to report from the meal just polished off than a vague sense of satisfaction. The belly could just as easily have felt bloated from the inundation of food or sick from indigestion. Throughout the alimentary journey—from stomach to intestines—eyes, nose and taste buds are conspicuous by their absence. Digestion does not need the approval of the senses. It works at a steady pace, without so much as a by-your-leave, under the direction of the autonomic nervous system. This system regulates everything you're too dumb to manage yourself: sweating in the heat, shivering in the cold, not falling from your bike, having erections on time, and emptying your stomach before the arrival of the next meal. Conscious supervision would only be counter-productive; it's best not to trouble yourself with these things, no matter how personal they are. Just trust your gut. No news is good news.

The only thing you have to do is gobble down the chow; as soon as it gets to your throat, your body takes over. Your body sees to it that the food is transferred into the energy you need to bring in more food. It's like being on a plane trip: once you've paid for the jet fuel you can leave it to the flight engineer and the captain to turn the fuel into energy, and to turn that energy into your journey. When the captain is speaking you don't even have to listen. Until the landing of the plane he has your fate firmly in his hands.

The process is reminiscent of pregnancy. Much of what takes place inside the expectant mother's belly occurs without her consultation. How to make a baby's head, or at what point the hands sprout fingers, or how many vertebrae go into the spinal column: it's all decided for her. The body knows just what to do, under the direction of the autonomic nervous system. The only time the mother plays a conscious role is at the beginning—mating—and at the end—giving birth. But giving birth isn't nearly as nice as mating. Here, we've got a bone to pick with the body. How about rewarding the woman during childbirth, too, if one reward—during mating—is enough to set the entire chain of cause and effect in motion and keep it going until the reproduction is a fact?

Fortunately, eating is rewarded twice: first at the table, then on the toilet. Apparently the body wants to be doubly sure that it gets rid of its burden. In childbirth the body doesn't have that much to worry about since the baby plays an active role. It wants to get out on time, before it gets too big and accidents start happening. For a baby, that's when life really starts. It's quite different in the case of a turd's birth. For a piece of excrement, excretion is almost certainly the end of the story. Excrement does not cooperate with its own downfall. In fact, at this point the excretory organs can really use a bit of encouragement from the body. An internal pat on the back, so to speak. Only then comes the satisfaction of delivery, as with a child.

The senses around the mouth are reminiscent of a front door and its paraphernalia: doorbell, name plate, mail slot, welcome mat, umbrella stand, all neatly polished and vacuumed. This is where you enter as the honoured guest. The back door is more appropriate for thieves in the night, making their stealthy escape. Nature has a good building code. The head faces forwards so it can see where it's going; the bottom faces backwards, where you don't want to know what you're leaving behind. While the head is very fussy about its incoming goods, the bottom drops its load and doesn't look back. There are some mighty weird animals in the world—sexual orifice next to the ear, ears on the knees—but no animal alive eats from the back and shits from the front. The strangest animal of all, the human being, spends billions of dollars on cosmetics for the front of the body, but there's no such thing as lipstick for the buttocks. 'Nobody lives out back,' as my mother used to say.

The enthusiasm with which you greet a meal at the front door does not have to be repeated when you bid it goodbye. The imported goods may have been screened down to the smallest detail, but the exported goods are released with barely a nod. There are no eyes or noses in the vicinity of the anus. You have to have eaten something terribly hot to experience a sensation that comes close to tasting when a turd is emitted. Generally speaking, the anus's lack of taste is seen as a blessing. So how do you reward it for getting rid of the body's rubbish at times and places we consider appropriate? The autonomic nervous system is not sensitive to pats on the back, so the central nervous system would have to be engaged. While that system may not have any idea what's happening to your food on the way down, your excrement breaks the silence at the end of the intestinal journey. You feel the need to go. 'The call of nature', as the English say, as if they themselves were unaware of it. At first the summons feels like a friendly tap on the shoulder, but woe be unto those who think it can be ignored. The friendly tap becomes an urgent tap, and then a smack. The discomfort increases until you think you're going to explode. Finally nature takes control, stops calling, and forces its way out.

The urge is inexorable. I have friends who have quit smoking, friends who give alcohol a miss, and friends who have terminated our friendship. But I don't have any friends who have stopped shitting. Nature's call is an order.

Pleasure is unthinkable without the senses. They tickle the nervous system when the body is being pleasantly stimulated, or evoke memories of some delightful tickle from the past: that sultry night with *her*, that cake from the baker who has since moved away, sweets at a children's party long ago. But if the back door has no sensory apparatus, how can you enjoy the exit? With an old trick. Instead of offering a reward, the body starts punishing. The punishment is gradually increased until it's practically unbearable. Then suddenly it's all over. What a relief! This kind of circuitous reward is called a catharsis. Filled with gratitude, you throw your arms around the torturer's neck—in this case your own body. But for what? Compared with the actual situation at the exit point you haven't gained a thing, but compared with the miserable situation your body has intentionally put you through you've gained a great deal. Your body makes you hungry, and thanks to hunger even raw beans taste good. But they're still raw beans. Happiness exists only in relation to unhappiness, as Sigmund Freud already knew in *Civilisation and Its Discontents*:

> One feels inclined to say that the intention that man should be 'happy' is not included in the plan of 'Creation'. What we call happiness in the strictest sense comes from the (preferably sudden) satisfaction of needs which have been dammed up to a high degree, and it is from its nature only possible as an episodic phenomenon. When any situation that is desired by the pleasure principle is prolonged, it only produces a feeling of mild contentment. We are so made that we can derive intense enjoyment only from a contrast and very little from a state of things.

No unloading without a load. Before defecation can take place, the intestines are pumped up by their contents like a balloon. There are no eyes to see the swelling or ears to hear it, but there are stretch receptors that warn the nervous system of the danger of bursting apart. When the nervous system gets this signal, it sends you an uncomfortable feeling. If the nervous system is favourably disposed towards you, it will search with increasing desperation on your behalf for an opportunity to reduce the pressure, which is constantly mounting. Is the meeting finally about to take a break, has the car finally turned off the highway, is the toilet finally free? Then the tension can be released—*pfffftt*—like a punctured balloon. Muscles relax, stretch receptors turn off their sirens, the guys in the emergency centre light up their cigarettes, and your entire being breathes a sigh of relief. This climax may be less fully articulated than the enjoyment derived from the meal that caused it, but it's no less inferior in terms of satisfaction. And as hard as it was to keep the load in, it's that much more delightful letting it go. Every now and then your efforts are

Catharsis

rewarded with a feeling of euphoria and ecstasy that under other circumstances can only be achieved after forty days of fasting, forty kilometres of running, or forty minutes of sex. A gorgeous turd stretches the colon so far that the vagus nerve goes wild, and your heartbeat and blood pressure drop in order to cut off the blood supply to the brain, which makes you high. *Mmm*! Although not every bowel movement can measure up to a run-of-the-mill orgasm or a tender Christmas turkey, shitting is among the simple pleasures that make life worthwhile—such as a drink before a meal, a grandchild on your lap, a grandpa under your bottom, and a well-orchestrated sunset. This pleasure is democratically divided: anyone can retreat to a sanitary confessional from time to time in order to reappear physically purified. Orgasms are not granted to all of us, but a life without defecation is inconceivable. Even the strictest celibate enjoys anal pleasure. As a child you're not the least bit interested in sex. Eating and pooing are distraction enough. Pooing unites necessity with pleasure better than eating does. That's why it's such a shame that so many people go to the loo with the same indifference that characterised Dutch eating habits in the fifties: just mash it up and wolf it down. Something seen as a necessary evil never has a chance to shine. Who would have thought back then that haute cuisine would gain a foothold in Holland, and that so many people would be eating more out of hedonism than hunger? You don't have to be a gourmet, however, to enjoy your daily meal, nor do you as an ordinary frequenter of the toilet have to have any particular specialities. Do what has to be done, but do it with a smile.

Nothing resembles faeces more than food. Both are organic substances that come in every gradation of tint and consistency: sometimes almost fluid, sometimes hard or fibrous, usually served up lukewarm. It isn't because of the material, however, that shit is held in such low esteem. What does it in for shit is mainly the lowly status of the organ in which it is housed, the intestines. In every living body, as in every business enterprise, there's a hierarchy of departments. Even though you can't do without a single one of them—except for a few, such as the appendix or a wisdom tooth—some organs are valued more highly than others. At the very top, literally, is the brain. In no other organ is the difference between us humans and other animals more evident in terms of size and complexity. High in the ivory tower of the skull, the brain receives only vague reports of what's going on down below, both in front and in back. Due to a lack of eyes, the control panel receives no images. Bright flashes and loud blasts of the siren are the only things that make their way through. It's only when this alarm is sounded that the intestines actually count, and only by calling in sick do the gall bladder and the pancreas get the attention they deserve. In many cases this is for the first and the last time, and you run aground due to lack of cooperation from organs you never even knew you had.

The more bestial a part of the body is, the less esteem we grant it. That's why so many ladies are glad they don't have tails, no matter how handy such a thing might be for chasing away flies. The mouth, on the other hand, is accentuated with lipstick. As an element of the face, it is among the dignitaries occupying the body's most chic neighbourhood, the head. Here the lips serve as a refined advertisement for the actual sex organs located elsewhere in the bestial hills. What happens there cannot tolerate the light of day. But this is more than compensated for by the brain at its most fanciful, as it drowns out the sloshing of sexual mucus with love songs and passionate poetry issuing from the highly-placed mouth. In the absence of such compensation, the booby prize goes to the lowest organ, the body's galley slave, the intestines. According to Plato, the head is separated from the torso by an isthmus, the neck, and the diaphragm divides the chest from the belly to prevent the lower region from besmirching the upper.

Hierarchy is a question of perspective. We regard our intestines as an auxiliary organ that provides the brain with energy. They're the stokers deep in the belly of the steamship over which the brain exercises authority. And that brain is us, according to Dick Swaab in his bestseller *We Are Our Brains*. Because our consciousness is located in our heads, that's also where our identity is. Unfortunately for our intestines, the nerves in the intestinal walls are less self-conscious. Otherwise we would have known better. It isn't the brain but the intestines that are the centre of life. Life is primarily the conversion of energy. In the intestines, energy from the outside world is converted into our living interior. Seen from the perspective of the gut, the sole purpose of all the other organs is to provide it with nutritious chunks of the outside world: hands to pluck them with, legs to reach them, teeth to grind them, a brain to know how to get at them. Of all those errand boys, what my intestines have the least regard for are brains. Can't stand them. Even though they've had to process their share of strange stuff. They've eaten rats, worms and beetles. Had to, for the TV shows. One director fed me sheep's balls, bull's cock and pig's cunt. His main purpose in filming me was to see how I would react. I cleaned my plate in every case. But I won't eat brains. I'd rather eat a shit sandwich.

What keeps me from eating brains is the texture. They're mushy. Mushier than most shit. It's like a limp handshake in your head. A slut of an oyster. I don't like oysters, either. I don't even dare shake hands with the kind of men who do slurp oysters. So what about the men with brains in their heads?

Why don't I like brains? And why do I like meat? There doesn't seem to be anything wrong with eating meat; all that muscle can only make you stronger. Spaniards eat bulls' balls for their potency. In certain restaurants you can even order the balls of some bull that has really outdone itself in the arena, by name and surname. Would I want to eat the brains of a particular individual? Einstein comes to mind. There have never been better brains than his. And they still exist. After Einstein's death in 1955, his brain was preserved in formaldehyde. There are photos of it. When you look at the photos it's hard not to feel disappointed. You'd expect a big carpenter to have a big hammer, but Einstein's brain is barely normal size, 1350 grams. After being weighed it was cut up into at least 240 little blocks. Choosing which block to eat wouldn't be easy. In which of the 240 is $E=mc^2$? That idea must still be in there somewhere. After all, every thought, every memory, is preserved as a set of connections between nerve cells, the way music is preserved in the grooves of a phonograph record. But you're no more capable of guessing what's in a brain just from looking at it than a man from Mars would understand that there's music in a CD or a whole story in a book. It's as if you had found a USB stick in a car with a forgotten police file on it and you had no playback equipment. The only playback equipment for a human brain is a human being. You'd have to transplant them. Oddly enough, the body would not think it had acquired a new brain—the brain would wonder how in the world it had ended up in that strange body, like a man realising he's sitting on the wrong bike.

Before we encountered our selves in our brains, we identified with our hearts. Civilised peoples like the ancient Egyptians believed that you thought with your heart. When they mummified their pharaohs, they threw the brain away and treated the heart with great solicitude. Modern Brits and Americans still don't know any better; they learn their examination material 'by heart', and not 'from their heads', as the Dutch do. In countries where Romance languages are spoken, the heart is invariably exalted above the head as the source of love. But there are other places as well where the heart is in charge of happiness. When cards are dealt, you always stand a chance of drawing a queen of hearts, but never a king of brains. There's no card game in which a jack of kidneys is played against the king of livers in order to trump the ace with a pancreas. We still love with all our heart. That's why we hate it so much when something goes wrong, and we're willing to spend so much money to have our heart and blood vessels repaired. Heart surgeons and blood specialists share in the honour that accrues to their favourite organs.

Gastroenterologists gnash their teeth. They know that the only purpose of blood vessels is transport. Real life takes place in the intestines. Here the substances from the environment around us are converted into the energy that makes life possible. For someone who studies the stomach and intestines, the heart and the blood vessels are mere intestinal accessories. There are oodles of lower animals that don't have hearts or brains, but an animal without intestines doesn't exist. Life began as intestines. The polyps in the ditch behind your house are nothing but tiny intestines on stalks. An intestine is all you need to be.

You are not your brain; you don't love with your heart; and even the horniest man is more than his dick. We are our intestines. At one time love may have been promoted from the heart to the brain, but as all cooks agree, love goes through the stomach.

Eating has status because it is social. It reinforces the bond between the eaters. Members of social species welcome any kind of behaviour that brings individuals together as a counterweight to the egoism that undermines the group's interests. The most well known means of strengthening group ties among people are religion and war. Eating is another.

Sex connects the best. You can't get any closer to each other than that. But there has to be an 'each other' involved. If you're all alone, the social bonus of sex is lost. The Bible says you must not spill your seed upon the ground because it does not serve the purpose of reproduction. But now that sex and reproduction have been disconnected for most people, solitary sex is not as highly rated. It isn't illegal, it no longer causes spinal tuberculosis, and most gods let masturbators into heaven, but you shouldn't be proud of it or talk about how great it was. In this regard it shows a remarkable similarity to that other source of solo satisfaction. After a pleasant dinner or a satisfying round of sex, it's customary to praise the performance of the cook or the partner to the skies, with accompanying facial expressions. But after leaving the bathroom, it's best to look as neutral as possible and to modestly maintain a stony silence about the pleasure your shit has just afforded you. It is of no social benefit. If you can't talk about it you can't share it, and unshared delight is no delight at all. How wonderful it would be to dish up your faecal adventures in all their fragrant and colourful glory! But you'll find no sympathetic ear at birthday parties for these particular feelings. It's as if you were talking about some family member who had made a bad investment, was still alive, and may be in need of discreet support, but about whom it's best to keep silent. People would rather talk about food and, after a couple of drinks, about sex. But as soon as the conversation switches to shit and piss it's time to be on your way.

Shit puts asunder what food has joined together. It's a perfect example of something you do outside the group. Yet shit also possesses an unexpectedly strong power to bind. It's true that shit drives people apart, but what makes it uniquely social is when you voluntarily set aside your repugnance of the distasteful stuff for the sake of another. Cleaning up someone else's shit creates a bond as strong as a secret pact. It's intimate material; shit is the currency of the love trade. Underlying this altruistic behaviour is the care of your children. Shit is always part of the picture. Your first thought, that people love their babies despite the stench of their nappies, is belied by the first pair of fresh young parents you see in action. The care they give to their baby is not in spite of the poo but thanks to it. Fathers and mothers do honour to their parenthood with offerings of poo and piss. After overcoming their initial revulsion, they actually grope around in it, greedily inhaling its odour and showing their unconcealed pride in an especially successful quality or quantity. On all baby poo and toddler muck, all snot from all little wiped noses, all drool and goop, love thrives like a rose on hog shit. It's precisely because it stinks and slithers that faeces can help you show how much you love the little pooper, how much nothing—not even poo—can come between the two of you. On the contrary, never do you feel so deeply connected as when the moist warmth permeates the nappy and your trousers and makes its way to your lap. As if it were your own.

While there's always a certain embarrassment involved in buying toilet paper for yourself, young parents—or old grandpas—on the checkout line at the supermarket flaunt the gigantic packages of nappies they'll soon be throwing away with their load of baby shit. A kinship of shit is a stronger confirmation of family unity than blood. While mother and child also share the experience of childbirth and nursing, shit is really the main means of connection between fathers and their children.

If you pay close attention you see that shit and healthcare are inextricably linked. With the help of the bedpan, hospital nurses forge an intimate tie with their patients that they never could have achieved with a needle, washcloth or IV drip. It's their job to save your life, no more and no less, but cleaning up your diarrhoea is always an act of mercy. If shit isn't the best vehicle for expressing unconditional love it's certainly the cheapest. In a society where love for animals is more easily exhibited than love for one's fellow man, an excellent role has been reserved for animal poo. Cat litter boxes are cleaned with averted gaze but quiet dedication, and guinea pigs get a handful of love with every handful of wood shavings. Poo is a godsend for lovers of fish and fowl. What better way to express their love than by the cleaning of the cage or aquarium? Telling people how much you love your canary doesn't make much of an impression; giving your neon tetra a gentle pat is likely to do it in. Feeding is better. Every mouthful of feed is a mouthful of love. Tenderly you watch your pet eat. It's still a cheap gesture, though, like a stale sandwich for the ducks or a spoonful of aquarium feed for a few cents. Shit is best. Cleaning up excrement is the essence of every form of house-pet affection. Every turd that makes its appearance is an opportunity to offer yourself up, the ultimate form of love in which you put the interests of others above your own. The high point of this shit cult is walking the dog. Its pooing takes up more of your time than its eating. What connects owner and dog isn't the leash so much as the shit. You have to go out at ungodly hours, but at the same time you can let the whole neighbourhood see how much you're prepared to do for your best friend. Walking the dog has long been the most important form of outdoor recreation in our country; for many streetlights, providing illumination is only a sideline next to their main job as dog toilet. The intestines that are lovingly filled at home are emptied outside. Men who have never seen their own wives defecating are willing to stand in rain and wind to enjoy their dogs relieving themselves. But all too willingly they pick up the warm and steaming excrement in a plastic bag and throw it away, something they'd never dare do with their own turds. After the walk, many a dog owner feels a sense of relief, as if he himself had just defecated. The dog curls up cosily in its basket. It knows it's done its owner a favour.

It even works with plants. Respectable women who discreetly cleanse their hind quarters with wet wipes dump huge amounts of manure on the roses in their garden, on condition that the gardener has brought them real cow shit. Plunging their arms in up to the elbows, they mix the shit with the soil and feel at one with nature. In this fusion, the smell of manure serves as incense. But the love that every garden flower or household dog gains from a little shit is not granted to everyone. Old people have a great deal of difficulty getting nursing care. Personnel are hard to come by, nursing homes are full, and many an elderly man or woman is given less attention than their cat. Yet when it comes to defecating they hold their own with their pets. Back when the old person was a baby he was passed from hand to hand and people willingly and lovingly cared for him, day and night; his bottom was washed till it shone and then royally powdered. What's the difference between a baby and a geriatric patient? Not much. Both drool, both have to be driven around, both know next to nothing about their own health, and both lie in their own filth. There's no use trying to give either one of them a good talking to; in both cases you'd be well advised not to wear your best suit when they press themselves against you. The biggest difference between a baby and an elderly person is the desire of others to care for them. Men who can't even imagine wiping the bottom of their old uncle poke around in the shit of their youngest offspring with a whistle on their lips.

Does old shit really smell worse than young shit? No, the shit is not to blame. It all has to do with the patient. People are born with an exceptional ability to elicit care. As babies they know just how to touch their parents' hearts by crowing, laughing and blowing bubbles. But that ability wears away, and by the time they reach adolescence there's almost nothing left. So most children leave their parental home, and not a day too late. Now they're on their own—until an accident puts them in the hospital. Or until they reach old age. At that point it's a good idea to activate the nurses' caregiving instinct. But how did you do that again, way back when? There are patients for whom nurses would gladly do anything because they don't moan and groan, they laugh. They have easy bowel movements. But being cheerful doesn't cut much ice if you're deathly ill or as old as Methuselah. Some patients wisely resort to acting like a baby. This is made easier by the nurses, who treat you like a baby anyway. It doesn't come naturally anymore, though. The ability to elicit care that you had in childhood no longer works. How could evolution have known that you'd still be a useful member of society in your old age?

How it all ends is a question of character. To elicit care, you have to live among a social species and be recognised as a member. Human beings are a social species, but they don't always regard everyone as fellow humans. While many people from human society are banished to nursing homes and institutions, non-human species like dogs and guinea pigs are taken into our hearts as honorary humans.

Shit plays a big role when it is being decided who does or does not belong. Michael Leahy, an Australian adventurer, found out how much shit connects people when in 1930 he went searching for gold in the most inaccessible parts of New Guinea, where the people lived as they had in the Stone Age and had never seen a white man before. Leahy and his group were received hospitably but with suspicion. When Leahy took off his hat, the natives next to him recoiled in terror. What kind of creatures were these? The most probable explanation, the tribe thought, was that they were dealing with reincarnated ancestors or other spirits. To find out for sure, they decided on an empirical test. Kirupano Eza'e reported as follows:

One member of the tribe hid himself and watched to see how they defecated. He came back and said, 'The men from heaven went over there to defecate.' As soon as they were gone, many men went to have a look. When they noticed that it stank, they said, 'They may have different skin, but their shit stinks just like ours.'

They were people just like themselves.

2

Nice and Nasty

Fire is hot, water is wet, God is great, the sea is salt, revenge is sweet. And shit? Shit is dirty. Through and through. Yuck.

If shit is dirty, I'm a dirty old man. All day long, hour after hour, year in and year out, I walk around with shit in my belly, warm and swaying like a foetus in a womb. And so do you. You and I, we're walking barrels of shit, chamber pots on legs. And under their tweed skirts and pleated trousers, the classiest ladies and gentlemen also carry intestines full of shit with them. No big deal. As long as it stays inside their tummies everything is hunky-dory. There everything is safely confined, like a bear in its cage, like Jack in his box, like Dracula in his coffin.

Shit isn't really dirty until later on, in the toilet. As soon as it comes out you don't dare lay a finger on it, something that even now is intimately shrouded in your delicate tissues. All of a sudden you don't want to have anything to do with it. No one to blame but itself. If only it had stayed inside, in splendid isolation. Who let it in, anyway? Not me. I wouldn't dare. It furtively gained access under the guise of food. Only after getting past the uvula, safely beyond reach of the sensory radar, did it throw off its cloak and show itself in the stomach for what it truly was: vomit. There in the interior it found itself in the abject company of snot, piss, mucus, sperm, pus and—its own future destination—shit. This whole mess would barely interest us if the body weren't so leaky. There are openings in it through which its contents meet our eyes, and our eyes aren't exactly overjoyed. How much more endearing is a person's view of himself in the mirror, no matter how ugly he may be, than this confrontation with what's inside!

Filth is only filthy when someone thinks it is—even if that someone is you. Halfway to the anus, one half of the turd has materialised while the other half is still pristine, split in two like a frontier village where the main street is also the national border. Only if the turd gets stuck does a moment occur when you wish you could take it all back, but by then it's already too late; as soon as the shit crosses the border it changes identity and becomes an undesirable alien. Once it's been pushed through, the internal half becomes external as well, and now, united with its better half into a single turd, it looks you in the eye. A vague glance of recognition passes between the two of you, but not until you experience 'the most brazen thing' that could await your eyes, according to the essay 'Shit' by Henk Hofland: 'the sight of the turd of your predecessor in the white porcelain bowl', 'the vilest thing of this type that could imprint itself on your retina'.

What could do no harm when it was still inside is suddenly suspect, only because it's now on the outside. In this regard, turd after turd obey the mantra of anthropologist Mary Douglas from her classic *Purity and Danger* (1966), which states that something is dirty only when it is out of place. As delicious as soup can be, and no matter how good a beard may look on you, soup is dirty when it's in your beard. Soup belongs in its bowl, snot in its nose, shit in its person. There's nothing wrong with the snot in my nose, no matter how slimy it may be. Nor can any fault be found with the snot in your nose. But your snot in my nose—now that's disgusting. A place for everything and everything in its place, lest the order be disturbed. This is all the truer for things that were already low in our regard. Excrement is expected to know its place, as are empty cans, old shoes, and certainly the lower classes, to keep them from pulling the higher classes down with them.

Saliva, for example, belongs in your mouth, where it performs all sorts of odd jobs in the process of digestion. Outside the mouth it's taboo. 'Think first of swallowing the saliva in your mouth, or do so,' the American psychologist Gordon Allport proposed in 1955. 'Then imagine expectorating it into a tumbler and drinking it! What seemed natural and "mine" suddenly becomes disgusting and alien.'

Any excrement encountered outside the body, like spit outside the mouth or extra-nasal snot, is a fugitive, illegal, on the run, bent on our destruction. We're up shit creek without a paddle! Thankfully shit doesn't have teeth or claws, so it doesn't seem to pose any danger. That's why shit doesn't make us anxious. People are not afraid of their own excrement. You don't call the police to deal with faeces (maybe you call the cleaning service). What you feel when you see a turd is not fear but disgust. You don't flee from a turd, you recoil from it. You don't look as if you've seen the turd but as if you've stuck it in your mouth. According to Charles Darwin, father of the theory of evolution and the study of emotional expressions, this is simply logical:

> As the sensation of disgust primarily arises in connection with the act of eating or tasting it, it is natural that its expression should consist chiefly in movements round the mouth. But as disgust also causes annoyance, it is generally accompanied by a frown, and often by gestures as if to push away or to guard oneself against the offensive object. With respect to the face, moderate disgust is

exhibited in various ways; by the mouth being widely opened, as if to let an offensive morsel drop out; by spitting; by blowing out of the protruded lips; or by a sound as of clearing the throat. Extreme disgust is expressed by movements round the mouth identical with those preparatory to the act of vomiting. The mouth is opened widely, with the upper lip strongly retracted, which wrinkles the sides of the nose, and with the lower lip protruded and everted as much as possible. The latter movement requires the contraction of the muscles which draw downwards the corners of the mouth.

By wrinkling the nose, retracting the upper lip, and screwing up the eyes, the innermost self isolates itself from the outside world. With this face, it's easy to tell disgust from the five other emotions that have their own facial expressions: sorrow, anger, surprise, fear and happiness. All six are not acquired but innate; you see them everywhere. Darwin found the expression of great disgust in the most remote corner of his world.

In Tierra del Fuego a native touched with his finger some cold preserved meat which I was eating at our bivouac, and plainly showed utter disgust at its softness; whilst I felt utter disgust at my food being touched by a naked savage, though his hands did not appear to be dirty.

A century later, the existence of universal emotions was confirmed by the travels of the Austrian biologist Irenaus Eibl-Eibesfeldt and the American psychologist Paul Ekman. Distant races of people could easily identify the emotions on photos of people far away. And blind people everywhere seem to make the same faces to express the same emotions. So facial expressions are genetic. But when it comes to disgust, there's something quite remarkable going on. Small children are simply unaware of it. Give them something bitter to eat and you get the expected facial expression, as sure as a vending machine will cough out a snack when you drop in a few coins. But they shove poo in their mouths with a smile. They vastly prefer one type of food to another, but they are not disgusted by filth. Parents exploit this by feeding them commercial baby food that they themselves couldn't imagine eating. The other five facial expressions appear much earlier. A baby shows happiness or sadness right from the start, followed after a few months by anger, fear and surprise, but it doesn't learn disgust until about three years of age. Apparently repugnance needs a part of the brain that does not develop until later on in order to express the innate ability to feel disgust. That area has now been identified, with the help of adults who, like young toddlers, can recognise all the emotions except disgust. These people are suffering from Huntington's disease, a form of physical deterioration caused by the death of the basal ganglia, a system consisting of four parts of the brain at the base of the cerebellum next to the insula, part of the temporal lobe. The insula controls things like sensualism, sexual pleasure, addiction and disgust, but also nausea and vomiting. The inability to read disgust on another person's face is often the first sign of Huntington's disease, which is marked by the failure of the parts of the brain that were last to develop when the patient was a toddler.

Because toddlers initially have no sense of disgust they're not easy to toilet-train. Fortunately the penny drops by the time they reach the age of three or so. At that point it becomes evident that there really is a hereditary aptitude for toilet-training. It's much easier to teach toddlers and pre-schoolers that shit or a dead mouse is dirty, while cake or a hamburger isn't. As parents, it helps if you screw up your face at the sight of shit or mouse, certainly if you say 'ewww' or 'yuck' at the same time—words that almost automatically make you wrinkle up your nose and put your mouth in spitting position. As off-putting as such a face may seem, it's also quite benevolent. Anyone beyond toddler age who sees your grimace will know they'd be wise to stay away from the food that your sense of taste has rejected. Why are the youngest children, the most vulnerable members of such a group, unaware of disgust? Maybe it's just as well that they have a close relationship with dirt, since it strengthens their immune system before they enter the highly sterilised world of modern adults.

Turning up your nose warns those around you that the food is bad. It's not a watertight system, however. What's disgusting to one person is delicious to another. And everything that tastes disgusting is certainly not bad for you, nor does all bad food taste disgusting. A lot of unhealthy food—hamburgers, salted peanuts, double whiskies—arouses desire rather than aversion. Apparently it hasn't been available in unlimited quantities long enough to horrify our genes. The consumption of such food causes a blissful expression that only makes onlookers salivate.

This doesn't happen with shit. In order to arouse repulsion, shit has a powerful card up its sleeve. From far away, long before you run the risk of actually tasting it, your nose recoils in disgust by way of warning. Shit stinks. Like shit. The smell is so unmistakable that you'd think it was an alarm signal, much like the smell they add to natural gas to make it smell like gas. You turn your nose away out of revulsion, followed by your head, if not your whole body. There's nothing wrong with the smell itself. Dung flies seek it out avidly, and many tropical plants imitate it to attract pollinators. But as a warning sign it cannot be ignored. It can even nauseate you. In this regard, smelling is actually tasting in a way: real faecal particles travel way up your nose and merge closely with the warm mucous membrane.

Brown, slimy and steaming with the stench, a turd is one of the most revolting things on earth. Perhaps a few improvements might make it more repellent—a bit of wiggling, sudden jerking movements, bilious green glow in the dark from the corner of your eye—but these things arouse fear more than disgust. A turd is not out to scare the living daylights out of people; inciting revulsion is enough. To this end, the turd has a whole arsenal of measures at its disposal. All our sensory organs come under fire. The ear, otherwise scaremonger extraordinaire, is spared the most, although it doesn't take much to imagine how a turd would sound if it were a bit noisier—witness the *Shlaarg*! and *Shlooderr*! with which dogs in comic strips produce turds for their owners to slip on. Sound is mainly good for creating unrest or causing irritation. The buzzing of a mosquito is irritating but it doesn't evoke disgust, and if the sound of farting were disgusting in and of itself no one would ever have invented the tuba. Generally speaking, the only time a sound revolts you is when you picture the corresponding action—puking, pooing or having an orgasm.

Compared with rich odour and full flavour, the voice of a turd is as quiet as the grave. Only the sense of touch comes under stronger attack than that of smell or taste. At least that's the impression you get from the language involved. According to *The Anatomy of Disgust* by lawyer William Miller, most words associated with disgust have to do with touch. While nasty tastes rarely go beyond bitter or rancid, and stench also has to be satisfied with a limited vocabulary, there's a wide range of ideas to choose from for describing disgusting experiences of touch: squishy, clammy, sticky, pitted, slimy and flaccid. In languages like English, repugnance is often expressed with 'sl' sounds: slimy slobbery slugs slithered through the slough like sleazy sluts. Words that start with 'sl' are usually up to no good. Just like slime and sludge, a slug consists of more or less stable material with lots of goo around it, along with the unexpectedly firm softness that you also find in the drooling lips of certain breeds of dogs and the little flaps surrounding those otherwise quite interesting sexual openings.

It's the indistinctness of a disgusting thing that makes touching it so exciting. The fact that it's neither stable nor fluid, neither dry nor wet but clammy, neither hard nor soft but flaccid, suggests that it doesn't know its place. A fresh cow pat is less substantial than an old dried-up one, which has taken on all the contours of a thing. You can't get a grip on fresh muck; it slithers through your fingers. But the worst is the 'neither warm nor cold' category. Tepid. Sitting on a cold toilet seat is not pleasant, and you risk burning your bum on a hot seat, but what's really awful is a seat that's lukewarm. Another person's bottom has just sat there. A living bottom.

Lukewarm is life. Miller points out that nothing arouses more disgust within the entire life spectrum than temperature. Anything that is as lukewarm as your own body alerts you to the presence of another life—on you, under you, or in you. Every other form of life is a potential threat, especially if it's unashamedly slimy and dripping. Inorganic things—rocks, plastic, brass—rarely incite disgust, but anything organic seems to want something from you, whether it's still alive or already dead. An atomic bomb is not disgusting, while the man who dropped it is. We're revolted by indefinable life out of fear of certain death. And then there's that shapelessness! Any respectable animal has a head (you can see its respectability in its eyes), and a good, honest plant has a stem with a flower. But a slug escapes definition.

Nothing so closely resembles a slug as a turd. Although it has neither head nor limbs, there's something in a turd that makes it seem alive. It's only apparently dead, like a vampire or a tulip bulb. Sometimes you may even think you've seen it move. It wouldn't surprise you at all if tomorrow you were to find it in a different place from where you left it. Legless yet mobile: now that's really something for such a lukewarm extrusion.

And so the eye proves to be the ideal sensory organ for registering disgust. Our eyes stimulate the imagination like the undeniable two-eyed animals we are. Even if you've never picked up or eaten a turd before, you only have to bring the image to mind to make your stomach turn. Imagining yourself stepping on a turd or a slug is enough to make you jump as if it were real. Because your eyes can see the shape of a turd but not the shit from which it apparently was made, most people would refuse to eat it, even if you've let the cat out of the bag and told them it was really cake. If you were to hold your nose it would be easier to eat a turd in the shape of a muffin than a muffin in the shape of a turd. Thus the people involved in a test conducted by American psychologist Paul Rozin refused to drink from a glass that had held a cockroach, even though the glass had been thoroughly washed and sterilised. It's easy to prove that this is a matter of disgust: little children cheerfully eat the muffin turd, followed by a nice glass of cockroach water. Their brains have yet to develop the imagination by which they later, as adults, will restrain themselves. A splendid example of the strength of the adult mind is recorded by Hermanus Hartogh Heys van Zouteveen, the Dutch translator of the work of Charles Darwin:

> In a zoo in the Netherlands, a young giraffe had broken its leg and was therefore put down. The director of the zoo sent a piece of meat from this giraffe to a family he knew well, specifying that it was a piece of meat from a deer from Baron V.'s deer park. A few days later he paid the family a visit and asked them how the venison had tasted. 'Absolutely delicious,' they said. The lady companion of the family had found the venison very tender and was inexhaustible in her praise. Now the director of the zoo told them that it really wasn't venison after all, but the meat of a giraffe. 'What, that big, yellow animal?' the lady companion cried out, and was struck by a violent attack of vomiting.

The image is stronger than the reality. If you don't know that someone has spat in your food, you don't taste it. The waiter who does it gets the usual tip. But if someone makes the claim, even if it isn't true, you push your plate aside in disgust. In times of hunger, this was a tried and tested way of getting your hands on your neighbour's food. According to Paul Rozin, things that have come from bodily orifices work very well: snot, drool, sperm, the remains of blood from a used tampon. But shit is always the best of them all. Shitting on a plate or eating out of a bedpan is the worst thing that can happen to you, certainly if it's someone else's shit or someone else's bedpan. If you're interested in a less lethal but equally effective variant of biological warfare, with a deterrent that's cheap and universally available, shit is the obvious choice. It's as if it were made for this purpose.

And it is. For a turd, being filthy is its ruling passion, the very point of its existence. The filthier the better, for it and for us. The filthier the turd, the greater the relief when you've got rid of it. A burden has fallen from your shoulders. But shitting is more than just a means of purging yourself of your waste. It's also a goal. Far more than merely a negative pleasure, defecating is a positive act: with every turd that is ejected your body is left cleaner and purer. More your own.

You can't be any cleaner than waste-free, and cleaner is better. It feels better, too. You look back on a successful session in the loo with the same satisfaction that you get from emptying the vacuum cleaner bag. The more you leave behind, the better the new person you become. There must be a market for a toilet with a built-in scale. Finally you'd be able to quantify your pleasure: another 185 grams cleaner! You could make do with a rough estimation, but that itself can be satisfying. You get the same surprise from an unexpectedly hefty haul from your ear or your nose, which makes nose-picking more popular than fishing. While fewer than 10 per cent of all Dutch people ever throw their line in the water, 91 per cent of the population regard their noses as rich fishing grounds, according to the national nose-picking test featured in the magazine *GezondNU*. Nose-picking can be tricky in public, but sitting in the car during rush hour and estimating the diameter of the little ball you've just rolled is a national pastime. And there isn't a single healthy individual who doesn't check to see what's under their fingernail after having stuck it in their ear. There used to be special little spoons for this purpose, but they've been banned by the medical profession after having caused too many accidents. It's better to have your ears syringed, not only because of the safety factor but also because it's a unique opportunity to proudly share the results. Any good GP will let you peek into the receptacle to see what your ear has yielded, and you leave the doctor's office with both ear and heart substantially lighter. It's equally satisfying for the doctor. 'Syringing the ear with lukewarm water to remove the mass within' is 'a rewarding task for the physician', according to the hygienic advice found in the pages of *Gezond blijven* (1929). Even the nose was syringed back then. To prevent fluid from ending up in the Eustachian tube, the patient was not allowed to swallow during the irrigation procedure. The best results were derived by having him stick out his tongue.

All this messing about with ears and noses pales in comparison with the cleansing of the intestines. For thousands of years, people have been applying syringe to anus: the clyster or enema. The ancient Egyptians learned the art from their god Osiris, who had cribbed the idea from the ibis. As enviable as a cat can be as it licks between its hind legs with its delicious little tongue, that's how dexterous an ibis is, reaching its nether regions with its long, curved beak full of water. In the mediaeval encyclopaedia *Der naturen bloeme*, Jacob van Maerlant reported the following with great admiration:

> When the ibis cannot shit,
> He takes water into his beak
> And ever so gracefully
> Purges himself below.

The Egyptians declared the ibis holy, and they imitated it. In the absence of a beak they cut off a length of hollow reed in order to divert the Nile and let it ripple sluggishly through their intestines. Specially selected doctors, the 'Shepherds of the Anus', administered medicines to them rectally. These doctors believed they could reach all the remote corners of the body via the rectum. Even blindness and toothache were attacked from that orifice. It would be centuries before the Swiss anatomist Caspar Bauhin (1560–1624) discovered a valve between the small and large intestine that blocks anything flowing in an unnatural direction. But this didn't prevent later doctors from challenging even death itself by anal means. This meant, for example, that a drowning person could be resuscitated with the help of tobacco. 'Take two full pipes of tobacco,' wrote physician Johann Gottlieb Schäffer in 1757, 'and light them. Put one in his bodily orifice and take the other in your mouth. Press the heads of the two pipes together and blow the smoke to the designated place.' 'However,' added Johann Georg Krünitz in *de Ökonomisch-technologische Encyklopädie* of 1787, 'one must take care that the pipe in the backside does not break off.' It could take an hour to smoke out the more obstinate life spirits. Out of breath, the doctor would resort to the tobacco smoke enema. Many a sea rescue society still has one of these among its old paraphernalia.

Before the development of the syringe, the doctor simply used a funnel to clean out your intestines (1556).

The enema enjoyed its greatest triumph when bloodletting was at its peak. Both procedures were based on the same principle: getting the bad out. You flush the sorrow out of your body with tears, and in the same way you got rid of ailments by draining off the dirty blood or intestinal water. Today it's just the other way around. In the hospital they no longer take the blood out of you but put it into you. If people from the eighteenth century were to witness a modern blood transfusion, they'd think it was the donor's health that was being treated. Likewise, concern for healthy shitting has shifted to concern for healthy eating. As today's thinking goes, the idea isn't to get a bad thing out of the body as much as to put a good thing into it: fibre, vitamins, plasma, an injection. All the better when you learn that the hypodermic needle originated with the enema syringe.

The actual enema syringe was invented in the sixteenth century. Before then, people made do with their mouth, a cow's bladder or a funnel. But even when real syringes were added to the arsenal it would take one or two centuries before the shit came squirting out of the intestines of Europe as if it were an Augean stable. First, however, there was one more obstacle to be removed: the practitioner. Administering an enema, like shitting, is something preferably done solo. Not only because of the shame involved, but also because of the clumsiness of so many domestic servants. As we read in Adrien Philippe, historian of the pharmacist's trade, it took a long time to get the hang of it:

> Now suddenly the barely practised hand begins to tremble. He searches but fails to find, hesitates, and needlessly exhausts himself. Then he sits down again, becomes entangled, and goes the wrong way; sometimes he is too lively, too enthusiastic, and is unable either to slow down or to stop; in other cases he is too frightened or too slow, and all he does is run skirmishes without daring to attack the target. Then he swings the weapon in sundry directions and touches that which should never be touched; or the hydraulic freight escapes through unobserved fissures and spurts over all the furniture in the room like wet rockets.

Those who were eager to take the business in hand themselves would have to wait for the improved enema syringe developed by Reinier de Graaf

(1641–1673), today better known for the Graafian follicles in which a woman's egg cells reach maturity. Thanks to a long, flexible tube between the pump and the mouth of the syringe it was possible to service your own backside from the front. This did not make the pharmacists happy. Now that anyone could do the job, they lost a chore whose distastefulness was more than compensated for by its remuneration. No doubt there also were pharmacists who took pleasure in it. But no sense complaining, De Graaf thought:

> The loss that carrying out fewer treatments might entail is balanced by the increased demand for the substance needed for the enema, since the doctors would prescribe it more frequently and the patients, no longer afraid to show their bottoms to strangers, would be more willing to use it.

With Reinier de Graaf's enema syringe you were finally in charge of your own bottom (1668).

In any case the path was cleared for the orgy of buttocks and syringes that took place in the seventeenth and eighteenth centuries. The French court set the example. Among the courtiers of Louis XIV the syringes out-spouted the fountains. Even the Sun King had himself cleansed two thousand times before giving up hope of ever getting rid of his painful bouts of colic. Nor did the enema have any effect on melancholy, night sweats, tumours or shortness of breath, but that did nothing to hinder its popularity, much like that other old panacea, 'bloodletting', and today's panacea, 'sports'. Under French influence, the syringe conquered Holland,

England, and the rest of Europe. Only the Germans refrained. The very thought of it made them blush to the roots of their hair, Krünitz wrote:

> The great repugnance that all true, decent Germans harbour against enemas can perhaps be explained by their custom of facing each other eye to eye, fist to fist, and to sincerely loathe all dealings that have the least appearance of backstabbing or underhandedness.

Enemas became fashionable, especially among the ladies, who saw them as a means of rejuvenation. Now that the Sun King was having himself cleansed, even during affairs of state and with everyone looking on, the most prim and proper ladies lost their inhibitions. They would have a lady's maid or girlfriend administer a cure when company was present, and the fact that people didn't wear underpants back then made it that much easier. With the shameless enthusiasm that modern women have for Botox treatments, these earlier sisters hoped to syringe themselves to beauty. Thanks to the prevailing fashion you could buy enema solution in all fragrances and colours, or exchange them with your friends. The syringes themselves, with their gilded silver, mother-of-pearl or inlaid tortoiseshell, were proudly displayed on dressing tables. And, like the plastic surgeons of today, few physicians had a critical word to say about it. The doctors themselves became the butt of derision, however, led by the French playwright Molière. He poked fun at them in his *L'amour médecin* (1665), *Le médecin malgré lui* (1666), *Monsieur de Pourceaugnac* (1669) and *Le malade imaginaire* (1673). While Monsieur de Pourceaugnac fled from a pharmacist who was bearing a gigantic syringe, *The Imaginary Illness* originally ended with a ballet involving eight enema syringe bearers, six pharmacists and twenty-two doctors. There followed an outburst of cartoons from Holland full of syringe-happy physicians, partly meant to decry the current political abuses but also to depict ladies with their elegant bare bottoms being joyously impaled with enema syringes by well-dressed gentlemen.

Satire is more effective against peevishness than against outright nonsense. Enemas are back (though they never really left). With the help of fasting, juice cures and laxatives, people are still detoxing in search of health. 'Detox' is a new word for the old idea that shit becomes poisonous if you don't dispel the last particle from your colon in a timely fashion. The ideal is to give your intestines a good purge from time to time, rather like spring cleaning: polyps on the table, baseboard and cabinets thoroughly scrubbed. In Great Britain, 5600 people a month have themselves cleansed by colon hydrotherapists; in the Netherlands there are dozens of colon therapy clinics, recommended as *the* way to get fit, to achieve emotional and spiritual balance, and to fight off numerous illnesses. Dozens of litres of water are pumped in through one tube and carried away by another. You can do it at home as well; in the Netherlands you're given a sieve so that when you're finished you can see what came out and what didn't. American doctors associated with *The Journal of Family Practice* are less enthusiastic. Among the serious side effects they mention are intestinal cramps, diarrhoea, upset salt balance, kidney failure, intestinal perforations and infections. There have also been deaths. So while there is little scientific reason to scrub out your bowels, the idea of beginning with a clean slate every now and then seems irresistible.

Cleaning is nature's obsession. Not only do humans spend the whole blessed day grooming themselves, but animals do too. Cats blissfully lick every delectable little cranny. Houseflies, which we dismiss as filthy, never stop polishing their wings and antennae. Big fish let little fish and crustaceans do their housecleaning for them by giving them access to their gills. The most important activity in the natural world is not eating or being eaten, nor is it mating and dying. It's grooming. For many animals, being their own cleaning lady is what it's all about. Grooming is usually more time-consuming than any other occupation. And rightly so: the tiny critters living in your own filth will do you in faster than any big predator from the outside world. That danger never lets up.

Dirt is tenacious. It obeys the law of conservation of matter more than any other material: nothing disappears into the void just like that. Usually cleaning simply means moving dirt around. From a hygienic point of view you'd be better off sniffing than blowing your snot into a handkerchief. In your trouser pocket the snot calmly keeps on brewing; in your stomach it's rendered completely harmless. Dogs and cats keep themselves cleaner by licking than we do with our combs and brushes. To us, licking is dirty, especially if our pets try getting us involved. Sloosh! One lick from a Great Dane and it's as if you had just come out of the shower. But clean feels different. There's enough spit on one dog's tongue for a thousand postage stamps and a hundred envelopes. That snout has just poked its way into a fresh turd. You feel sorry for an animal that has to walk around all day with such a filthy length of muscle in its mouth. Yet you often hold something quite similar in your hands. Just as big and slippery, just as supple, and just as filthy as the dog's tongue. There's no dog connected to it, however; it's a tongue that's in business for itself. We know it as 'the rag'. You use it to wipe down the table, clean out the gutters, and dab out blood and egg stains. It makes things clean and dirty at the same time. As you wipe down that table, you're introducing bacilli from the last job in expectation of a third, when you'll rub in the filth from the previous two chores. Rags are to bacilli what aeroplanes are to us: they close the gap between two distant points. A bacterium can swim short distances by itself, as long as its watery world doesn't dry out. Fortunately, a tidy housewife usually comes along with a wet rag to help her little housemates continue on their way.

Grease goes from your hands to your napkin, saliva from the speaker to the audience, shit from the intestines to the toilet. Dirt here becomes dirt there, dirt there becomes dirt here. And dirt that doesn't move on just stays put. Dirt is invincible, thanks to its adhesive strength. That's what makes dirt so dirty. It stays where it is, and if you succeed in getting rid of it, it sticks to whatever you used to remove it. The worst disgust you can feel is with dirt that sticks to you and thereby becomes part of you. Those who handle pitch are easily soiled.

Dirt on your body inspires fear. You can't run away from it. It isn't an enemy that you can attack because it's on or in your body. It has taken possession of you. William Miller uses the concept of 'horror' to describe such fear-filled repugnance. As in a horror film, there's no escape. The evil lodges itself in you, and when it comes out it doesn't let you go.

Shit is right at home in the body. Shit forms inside you like a malevolent demon that won't let itself be driven out without leaving traces behind. The turd splashes into the toilet on its way to the caverns of the sewer, but there are always little lumps that remain, brown stuff around the anus, in the crack between the buttocks, latched onto the hairs above the deep abyss like true survivors. Sometimes you can feel a speck of shit, broken off the last turd and hanging from your arse like a little tail, too light to mean anything to the force of gravity, ignoring the futile squeezing of the sphincter, a disgusting barnacle. So there you are, stuck, half upright, your trousers around your ankles, your backside too filthy to pull them up. You've soiled your own outside with your own inside. There's no turning back. Nothing for it but to clean yourself up. But how do you get rid of all that goo? Lick it off?

Humans aren't limber enough for that. Most Westerners make do with paper instead of a tongue. It's an unequal fight. Paper is virtually powerless when it comes to shit's sticking power. With every wipe the paper seizes remaining bits of poo, only to be forced to surrender them to the hairs of the crack in your butt. One sheet of paper isn't enough, if only because you might perforate it with your finger. But even after many sheets, you wouldn't dare use the last one to wipe your lips. Manufacturers laugh up their sleeves. By taking advantage of our fear of our own filth they've enslaved us to something we can never get enough of: one more sheet, then one more. Only after eight to ten sheets does the average user dare to re-engage with life beyond the bathroom door. So with a family it doesn't take much to go through a roll of toilet paper. That explains the success of Hans Klenk, whose 'Hakle 1000-Blatt-Rolle' made him a leader in the German toilet paper market, at 500 million euros a year. Yet specialised toilet paper hasn't been around all that long. It appeared at more or less the same time as sewer systems, at the end of the nineteenth century. 'The original and only genuine Gayetty's Medicated Paper' (patented 1871) was recommended as a treatment for haemorrhoids. Gayetty claimed that haemorrhoids were caused by wiping with ordinary paper, which had become available with the distribution of daily newspapers. From my own youth I distinctly remember the old newspapers neatly cut into sheets and hung from a wire on the door of the 'little house', below the cut-out heart. Pages from books were also used. Lord Chesterfield (1694–1773) advised his son in a letter to always carry a cheap edition of the Latin poets around with him. Then he would have something good to read on the toilet, and with a practical application for every page he read. The letter written by the German composer Max Reger to the critic Rudolf Louis in 1906 is in the same vein:

> I'm sitting in the smallest room in my house. I have your review in front of me. Soon I'll have it behind me.

In America, mail-order catalogues were popular as toilet paper until the 1930s, when the switch was made from highly absorbent matte paper to smooth, glossy paper. Sulking city dwellers now had to pay good money to wipe their arses. In the countryside many people remembered that you could get by without paper by using corn cobs, and full barrels of cobs were always close at hand. Out in the fields there were twigs, stones, leaves, bones or handfuls of grass. Monks in the Middle Ages used old rags or potsherds. But for centuries, even those in opulent circles never considered wasting something as costly as paper. At the French court, Madame de Maintenon, the second wife of Louis XIV, used sheep's wool.

Cardinal Richelieu preferred hemp.

Today you're not likely to see red-faced individuals at campgrounds walking around with rolls of hemp or sheep's wool under their arms. It's got to be paper. But it still isn't something that can be openly discussed. In advertisements for toilet paper you never see turds; what you do see are teddy bears and fleecy clouds, as if the product being sold was a buttock softener and not a shit catcher. Modern people are accustomed to advertising nonsense, but in the sixteenth century the sight of little bears on a poster for bottom wipers would easily have given rise to confusion.

It's possible that François Rabelais would have taken it literally, as his *Gargantua and Pantagruel* testifies:

> I have, answered Gargantua, by a long and curious experience, found out a means to wipe my bum. Once I did wipe me with a lady's neckerchief, and after that I wiped me with some ear-pieces of hers made of crimson satin, but there was such a number of golden spangles in them (turdy round things, a pox take them) that they fetched away all the skin of my tail with a vengeance. This hurt I cured by wiping myself with a page's cap. Afterwards I wiped my tail with a hen, with a cock, with a pullet, with a calf's skin, with a hare, with a pigeon, with a cormorant, with an attorney's bag, with a montero, with a coif, with a falconer's lure.
>
> But, to conclude, I say and maintain, that there is none in the world comparable to the neck of a goose, that is well downed, if you hold her head betwixt your legs. And believe me therein upon mine honour, for you will thereby feel in your nockhole a most wonderful pleasure, both in regard of the softness of the said down and of the temperate heat of the goose, which is easily communicated to the bum-gut and the rest of the inwards, in so far as to come even to the regions of the heart and brains. And think not that the felicity of the heroes and demigods in the Elysian fields consisteth either in their asphodel, ambrosia, or nectar, as our old women here used to say; but in this, according to my judgment, that they wipe their tails with the neck of a goose, holding her head betwixt their legs.

Modern people are not likely to reach for a little bird or a teddy bear when they need toilet paper, but they're not told how to use the stuff either. It doesn't come with instructions. So people improvise. Staying seated while wiping is awkward because you can't reach what has to be reached. There's no hole in the front of the pot, as there were in the ancient Romans' public toilets. Many people today simply stand up. But in this position you automatically squeeze your buttocks together, smearing the shit from one buttock to the other. The best way is to lean forward so your buttocks open up obligingly to let themselves be pampered.

How do you pamper a buttock? That depends on the buttock. But certainly not with paper. Even a little bird or a teddy bear would be better than that! Anyone who has tried to clean greasy hands with a paper napkin during a posh dinner will understand that cleaning an arse with paper is doomed from the start. People don't wash their feet or dishes or brush their teeth with paper, but when it comes to the filthiest of all filthy places, the very abode of the devil on earth, paper is considered good enough. You and I, we wipe our bottoms with paper on the authority of our parents, who in turn learned it from their parents. But they never sat on clean buttocks either. I understand why people who fancy themselves clean put on fresh underpants every day. At least that helps a little in removing the brown bits that were first swept under the rug, as it were. Moving filth around in the loo is suspiciously like moving paper around at the office. It doesn't solve anything. To really get rid of filth you have to transform it until it's no longer recognisable as such.

This is best done with water. In the eastern and southern countries of the world that's common knowledge. There they wash their backsides the same way they wash their feet and dishes and teeth: with water. Every toilet has a bowl of water at hand, or everyone carries their own bottle with them. In countries where water is in short supply this is a practice readily accepted. But in countries like England and the Netherlands, full of canals and rivers, mist and fog, the tap is located on the other side of the bathroom door where you can't reach it when you need it, because that's where people are supposed to wash the hands that they soiled with the paper that got dirty without doing any cleaning. While people in the West fill the toilet with enormous amounts of water to flush the turd from sight, they can't set aside a little bowl of water to wash away its traces.

Water is virtually a universal solvent. Inorganic substances are broken down into their salts, the salts to their ions. Organic substances are quickly converted by the microbes that flourish there. Thus water removes practically all stains: the dirty dishes, a bad taste in your mouth, the sins of the world, and shit. In a soggy climate like that in the Netherlands, a turd outdoors disappears in just a few days. People wouldn't complain so much about dog droppings if so many new droppings weren't being made and if people didn't love complaining so much.

Yet water isn't what it used to be. These days, the symbol of purity itself is in need of a good washing. There are vast water purification plants outside every city. Natural purification by means of evaporation and precipitation can no longer keep up with the level of pollution. Before the fresh rain ends up in your glass of water, it's already gone through the retorts of many factories. And the intestines of many little creatures. On its way from retort to intestine to retort to intestine, water has had to swallow a wealth of dirty substances. And dirty substances are only the beginning. There are dirty microbes in it too. We knew that even before the invention of purification plants. Microbes were discovered back in the seventeenth century by Antonie van Leeuwenhoek, one of the first scientists to use a microscope. Water turned out to be full of 'animalcules', tiny animals. From then on you didn't have to travel long distances to see strange life forms; each drop of water contained an immense universe with countless bizarre little creatures, all within reach. Never before had anyone discovered something so tremendous. But it left people cold. For two centuries, nobody cared a fig about the tiny little 'animalcules'. Until 1876, when Robert Koch demonstrated that they could make you ill. Deathly ill. One after another, the causes of tuberculosis, cholera, typhus, diphtheria, tetanus, plague and syphilis were quickly unmasked. Suddenly we appeared to be besieged by a danger as immense as it was invisible. We've never gotten over the shock. In the words of medical columnist Lewis Thomas, we're still living

> in a world where the microbes are always trying to get at us, to tear us cell from cell, and we only stay alive and whole through diligence and fear.
>
> We still think of human disease as the work of an organised, modernised kind of demonology, in which the bacteria are the most visible and centrally placed of our adversaries.

Microscopic riffraff are bad company, and like all bad company they keep hanging around. If they don't cling to our bodies directly, they adhere to the dust we inhale, the drops we drink. Every coughing fit is a form of public transport to a virus; every bystander is a terminus or transfer station. As a cougher you really don't want this on your conscience. If you're properly behaved you cover your mouth with your hand. But the little travellers couldn't wish for anything better. All they have to do is wait for you to shake hands with someone else.

You can catch a cold from viruses in someone else's cough, and bacteria in drops of saliva cause tuberculosis, a disease from which ten thousand Dutch people died each year at the end of the nineteenth century. Shit can give you cholera. The bacteria in a turd are just as skilled at flying as those in a mouth. But with ten to a hundred billion bacteria per gram of shit, when the final splash comes there are enough of them spattering your bottom, moving on to your fingers after wiping, or sticking to the rim of the toilet to put your antimicrobial toilet block to shame. Washing your hands doesn't really help; before you even get that far you've dropped clusters of bacteria off on the handle of the flushing mechanism, ready for the next lift. If you really wanted to reduce the number of bacteria once and for all, you'd have to scrub your hands like a surgeon for minutes at a time with antiseptic soap without touching the tap. Nevertheless, most people—their hands full of new bacteria, having given the shit a good rub into the hair between their buttocks—are firmly convinced that they come out of the toilet cleaner than when they went in. The tidy Dutch never say they're going to empty their bowels but that they're going to wash their hands, and fastidious Americans call their toilet a 'bathroom'. Actually, the toilet is more and more frequently being placed in the bathroom. With every push of the flushing mechanism a cloud of faeces particles from the toilet can easily reach a distance of five metres—over to the sink, over to your toothbrush. Even so, the toilet is to be preferred to the kitchen. Kitchen sink and kitchen towel are paradises for bacteria. This is where all the rubbish from the entire kitchen come together. If I were a colleague from Mars, an American microbiologist once said, I'd shit in the kitchen sink and eat on the toilet. But it isn't at the kitchen counter that you run the most risk. An old-fashioned housewife is safer from bacteria than a modern career woman with her digital arsenal. In a four-year study in five major American cities it was shown that 550 pathogenic bacteria were present on a single computer keyboard, 400 times more than on a seat in a public toilet. According to an article in the *Journal of Hospital Infection* from 2006, you're much more likely to contract meningitis or swine flu by making a phone call, working on a computer, or withdrawing money than by licking the seat of a toilet in a public cinema. Mobile phones are particularly active breeding grounds. Fifty million bacteria have been counted on a single phone. Yet computers and mobile phones don't evoke anywhere near the disgust they deserve. They haven't been around long enough. Mothers don't yet teach their children that plastic is dirty, so it's way too early for such an aversion to be genetically ingrained.

For pious Christians, shit is a product of the Fall. Before evil entered the world there was nothing to defecate. God never did it, of course, not even in his human form as Jesus. In the Bible he ate and drank to his heart's content, but there's no mention of defecating. But what happens when someone eats Jesus in the form of the host, which has really changed into His Body? According to the heretical teachings of the Stercoranists, the consecrated host ended up as a turd, just like ordinary bread or meat; so people who went to Communion during the fast committed a sin because they weren't withholding from eating. Rome strongly condemned these ideas, which conflict with every form of devotion. No one found it necessary to test them by eating nothing but consecrated hosts for a week and seeing if any shit came out. Shitting God is simply an internal contradiction; good and evil do not go together. That's just what Milan Kundera thought in *The Unbearable Lightness of Being*:

> When I was small and would leaf through the Old Testament retold for children and illustrated in engravings by Gustave Dore, I saw the Lord God standing on a cloud. He was an old man with eyes, nose, and a long beard, and I would say to myself that if He had a mouth, He had to eat. And if He ate, He had intestines. But that thought always gave me a fright, because even though I come from a family that was not particularly religious, I felt the idea of a divine intestine to be sacrilegious.
>
> Spontaneously, without any theological training, I, a child, grasped the incompatibility of God and shit and thus came to question the basic thesis of Christian anthropology, namely, that man was created in God's image. Either/or: either man was created in God's image—and God has intestines!—or God lacks intestines and man is not like Him.

Martin Luther eagerly seized on shit's bad reputation to sing God's praises. 'To illustrate his own worthlessness and that of the world in relation to the greatness of God', he compared himself to 'ripe shit' and the world to 'a gigantic arse': 'For this reason we had better go our separate ways'. The smell of shit rises from all of Luther's work like a metaphor for everything that is of the devil, with the detested pope leading the way. 'The pope is a bishop of the devil and the devil himself, yes, the shit that the devil himself has shat in the church; that is the shit by which they say you will be blessed.' Shit was the obvious choice because Luther suffered from chronic constipation. His struggle in the loo was for him equivalent to the hopeless task of freeing himself from the evil within. But it gave him plenty of time to think. It was during one of his conflicts with his intestinal demons that the basic principle of the Reformation occurred to him— that we cannot expect to be saved by works but by faith. Luther made no secret of the source of his divine revelation: 'on the privy in the tower' of the Wittenberg monastery. How

The Lutherans of 1545 didn't give a crap for the pope. They knew just what to do with his tiara.

different the course of Western civilisation would have been, sighed shit expert Dave Praeger in 2007, 'if Luther had eaten more fibre'.

Throughout all of church history, material filth and spiritual filth are closely intertwined. It begins right away in the Old Testament, when God declares a large portion of His own creation to be unclean. Seldom had He been so inscrutable. In their attempts to explain the Old Testament laws of impurity concerning pigs, newly delivered mothers and excrement, centuries of scholars tried to out-rationalise each other. Pigs are useless in the desert, newly delivered mothers need their rest, and shit carries disease. In their divine wisdom, priests and prophets were supposedly providing sensible medical advice wrapped in pre-scientific religion. For believers, however, health was of minor importance. What they were interested in was the salvation of their immortal souls, not household tips. 'Even if some of Moses's dietary rules were hygienically beneficial,' writes Mary Douglas, 'it is a pity to treat him as an enlightened public health administrator, rather than as a spiritual leader.' The American philosopher William James called the custom of basing the religious concept of 'impurity' on hygienic motives as 'medical materialism': 'we kill germs, they ward off spirits'. But 'we' are still not very different from 'them'. Most people would refuse to accept the transplanted heart of a murderer, according to an article in *Harper's Magazine* (2009), even if their lives were at stake. Likewise, Paul Rozin's experimental subjects wouldn't think of putting on a sweater that had belonged to Adolf Hitler, although it had been washed and drycleaned. Apparently they had the idea that some ineradicable evil from the former owner was still adhering to it. How easy it would be for that evil to permeate your own body and work its way into your very soul.

If the evil is less tenacious then at least you can try to scrub it away. Even a child understands that. A great many children have had their mouths washed out with soap after having said certain dirty words. But adults also believe in spiritual cleansing with soap and water. According to the magazine *Science* of 8 September 2006, people who feel guilty are more inclined to wash their hands than those who don't. This is called the 'Macbeth effect' after the tragedy by Shakespeare in which Lady Macbeth desperately tries to get her hands clean after the murder of King Duncan. Although most people don't commit murder, everybody feels a little guilty about something—guilty enough to feel like a better person once you've washed your hands, brushed your teeth and relieved yourself of urine and excrement. It's as if you had washed your soul too.

Mario Vargas Llosa, in his erotic novel *In Praise of the Stepmother*, describes how Don Rigoberto felt after successfully relieving himself: 'there invaded him that intimate rejoicing at a duty fulfilled and a goal attained, that same feeling of spiritual cleanliness that had once upon a time possessed him as a schoolboy at La Recoleta, after he had confessed his sins and done the penance assigned him by the father confessor.' Nothing lends itself more readily to metaphorical use than dirt and washing. You don't need a mop to resolve dirty business. Smut can be got rid of without soap. It isn't the body you cleanse in the baptismal font as much as the soul. If the dirt is of the devil, then washing is a form of exorcism. In this way you can elevate not only yourself but your entire tribe as well. At least that's what the nineteenth-century hygienists thought. While the family doctor was concerned with healthy people, the hygienist was concerned with a healthy society. In *The Skin*, written in 1860, Dutch physician G. D. L. Huet's argument for washing the skin was all about body and spirit:

> Who among us has not discovered the stimulating and reviving effect that a fresh bath, an adequate cleansing of the body, can have on our spirits? How unpleasant one feels in a dirty suit and grubby skin? Yes—let me say it—how more or less ashamed one feels about oneself, while after washing this shame gives way to a heightened sense of self-respect.

Such arguments were not without self-interest. The upper classes felt physically threatened by the contagious diseases that bred in the slums. Tidiness and purity could only succeed with an appeal to conscience. The primary usefulness of light and fresh air was to make you feel like a cleaner person; actual cleanliness then followed of its own accord. But the upper classes weren't entirely sold on this plan. Even the socialist writer George Orwell took a gloomy view. No differences in education, upbringing, race or religion were great enough to shake his belief that all people are equal, yet physical revulsion for him was an insurmountable barrier. He wasn't bothered nearly as much by murderers or sodomites as he was by people who slurped their food. Travelling through the pre-war industrial areas of northern England in *The Road to Wigan Pier*, he confirms the bias of the European bourgeoisie—that they could never regard a labourer as their equal—with four terrible words: 'the lower classes smell'. Labourers would have to be cleaned before they could be uplifted. And so it was. The working class were scrubbed and polished up a rung, with tidy little gardens and clean bathrooms, but they in turn began looking down on the lower class from which they themselves had risen.

History repeats itself. In the classic *The Civilising Process* (1939), sociologist Norbert Elias shows how squalid manners have been tamed since the Middle Ages with the help of class distinction. Erasmus knew as early as 1530 that a snotty nose was not the done thing. A peasant blows his nose in his cap or clothing, a peddler wipes it on his arm or elbow. Using your hand and then rubbing it on your clothing isn't much nicer. 'It is proper to catch the filth of the nostrils in a handkerchief', Elias writes in his book of etiquette *On the Politeness of Children's Manners*. But it is not proper, adds the *Galateo* of Giovanni della Casa, archbishop of Benevento in 1558, to unfold the handkerchief after using it and 'to look inside as if pearls and rubies had fallen out of your brain'. In the absence of a handkerchief, Erasmus concedes, the hand may also be used, but 'as soon as the snot is cast to the floor, the nose having been blown with two fingers, one must immediately rub it out with the foot'. *La Civilité Françoise* of 1714 also allows you to wipe away spittle that has landed on the floor, as long as you never spit so far that you have to go searching for the spittle before you can put your foot on it.

The upper classes were careful to obey the rules so as not to be taken for members of a lower class, out of shame. You had to uphold your social position at all costs. And you couldn't do that with a dripping nose or with your trousers around your knees. If people see you like that you feel as if you'd been caught, you blush to the roots of your hair, and you want the earth to open up and swallow you whole. As the lower strata adopted the rules of the upper, shame was replaced by revulsion; one avoided squalid behaviour not because it might disgust other people but because it disgusted oneself. People are embarrassed if someone is disgusted by them, and they are disgusted if someone behaves disgracefully. Civilised people defecate in isolation mainly because they are disgusted by the beast that dwells within them that has to shit so urgently. But shame came first.

In the beginning was shame. It's in the Bible; check it out. Right in the first book, Genesis, after the Fall, describing Adam and Eve: 'And the eyes of them both were opened, and they knew that they were naked; and they sewed fig leaves together, and made themselves aprons.' They had each other to be ashamed in front of. But even when no one else is around to see you in the altogether, as Erasmus knew, you must expose yourself with the appropriate hesitation, for the angels are always present. There's nothing they'd rather see in children than hesitation, the companion and guard of modesty. While few modern people believe in angels anymore, the feeling that you're always being watched is still with us. It's even with the current icon of shamelessness, the German writer Charlotte Roche, who has shared her most intimate perversions with hundreds of thousands of readers in *Wetlands*. During her adolescence she exchanged tampons with her best friend Irene. From one red cunt to the other, from the other back to the first. 'Through our old, stinky blood, we were bound together like Old Shatterhand and Winnetou. Blood sisters.' But in an interview with *Vrij Nederland*, Charlotte Roche confessed that as a teenager while getting dressed in her room at home she would suddenly find herself fumbling. She was embarrassed in the presence of the boy bands who were looking down at her from the posters on her wall.

Shame is inextricably linked to civilisation. Every turd, every fart reminds a civilised person of their bestial origins. Animals themselves know no shame. They mate openly and naked, and shit with you looking on. Except for that one cat I heard about (and with great sympathy) who refused to use its litter box if the household dog was watching.

Shit is filthy. That's just the way it is. But shit is something you can't live without. To come to terms with your filthy excrement there are three paths open to you. First, you can tame your shit. Keep it out of sight, eliminate the stench, defecate furtively. This will cost you miles of toilet paper, thousands of toilet doors, stacks of etiquette books, handfuls of money. The second method goes one step further: deny everything with a brazen face. Act as if human beings didn't shit at all. Be silent as the grave about it. That's what the Victorians did. The fact that they invented modern sewer systems is no accident. Sewers enabled them to spirit every turd away from daily life; hygiene was merely a by-product. As long as everyone pretended to know nothing about it, there was nothing to know about. We still do this. Everyone knows what kinds of unsavoury things are taking place behind the doors neatly marked LADIES and GENTS, and everyone plays dumb. But with this method you might as well not have defecated at all. There's no fun in it.

Fortunately there's a third way: enjoy it. Take your time. Ignore the raised eyebrows when you spend more than fifteen minutes in the loo. Relax. Wait patiently until the pleasure makes its way from the lower to the upper regions. Don't let the smell offend you; it's only yours. Filthy is filthy, but—like sex—it can also be glorious. As a child you knew this without thinking about it. Splashing in the mud, pinching pimples by yourself or with a friend, eating your boogers, playing with your food, looking to see who has the longest turd—you hardly had any time left for wasting. Back when you were young, party stores lived up to their name, with their fake turds, artificial puke and stink bombs. One or two kids figured out how to sublimate the fun as adults. Kids who messed around with shit became sculptors; a few boys who liked to pull the legs off flies became biologists. Most of them have lost the knack, however. For years, their passion has been held in check by disgust. But there's no need for this. Because passion and disgust almost balance each other out, only a small push is necessary to tip the scales to the passion side. You can start on the toilet. Take it easy; don't push too hard.

Is anything coming?

3

Private Delights

People don't come in single units. They come in pairs, in teams or in hordes. Ladies come in reading clubs, old folks come by bus, soldiers by graveyard. We are a social species. Although the newspapers are full of anti-social behaviour—man hit on the head with a blunt instrument, woman raped—these are exceptions. Otherwise they wouldn't end up in the newspaper. You and I, we went another day without raping anyone, and our blunt instruments have never come out of the drawer. We put the rubbish bin out on the kerb and wished the man next door good morning. Because we cannot do otherwise.

Most animals aren't so stupid. They don't live social but solitary lives. A tiger doesn't need the company of another tiger until mating season. A real cat walks by itself, unlike wolves and dogs, who like to stick their noses in the bums of their fellows and vice versa. For them there's strength in numbers. But they, too, suffer from the curse of every social animal: never-ending conflict. They may work together to snatch a cake, but once that common cake has been snatched they fight each other to gain possession of the biggest piece. A solitary animal like a cat looks on such behaviour with disdain. It doesn't need a slice of cake; it would rather have a cake all its own.

Our species is social, alas. Like a stray horse along the roadside or a wolf without its pack, a human being is constantly yearning for fellow humans. If they're not at hand, loneliness threatens. You see them in the newspaper or on television, and, if you look carefully, behind the curtains in your own street: lonely people, waiting for the son who never comes to call, for a strolling cat in need of patting, for death. Deeply distressing. But the opposite can also assume unbearable proportions, and there isn't even a word for it. I can feel awfully *excessified* sometimes. After a whole day in groups—the family, school, work, card club—you can get fed up with your own species. Sometimes, in the middle of a party, I suddenly wonder what I'm doing there. Then someone will say *we* to me again. The *we* that suggests we're all in this together, the *we* that fortunately is not *them*—the *we* that they conclude I am a part of. The *we* of the nurse in the hospital. When you're lying there helpless in bed, and you know that tomorrow she's going to ask you again, 'So, Mr Dekkers, have *we* moved our bowels today?' *That* we. If ever there was a place where you'd rather be an *I* than a *we*, it's in a hospital room when you have to go. How many sick people blanch with the embarrassment they feel at such a time among their fellow humans? I'm a loner, personally. Rather than doing big things in a team, I prefer doing small things in a corner. But where? In this vast sea of humanity, where can you find a deserted island to recover your equanimity? Where they finally leave you alone?

On the toilet. There's privacy in the privy. You can withdraw into a public loo without anyone saying a word. A toilet is an escape capsule for fleeing from excessification. In solitary seclusion you find the rest you need for the one thing that has to be done there. If you have to be alone to shit, you have to shit to be alone. Here you can do what is strictly forbidden everywhere else: sit on your bare arse, stink to your heart's content, pull globs of poo from your bum hair, and—a mortal sin for any social animal—be by yourself for a few minutes. For this little space of time the individual is more than a part of the masses.

As a pretext for these privileges the toilet is designed as a more or less refined shit sanatorium. Over the centuries humanity has devoted to this effort the same abundance of ingenuity as it has to all those patented mousetraps, cherry pitters and nutcrackers. But all the toilet variants are based on the principle of the hole. Whether it's a pit in the ground, a platinum pot, a French squat toilet or the most modern Japanese facility, they all consist of a hole with something attached to it. The excrement comes out of your body via your arsehole and passes into oblivion via the hole in the toilet. You cover the pit over, toss out the contents of the pot, or flush the toilet. In all three cases the actual work is done by gravity. In a mediaeval castle, the turd fell out of the hole in the toilet, went between the walls of the tower and into the moat. Privileged citizens had an oriel—a garderobe—with a hole in it. Little had changed since our ape days when we just shat in the open whenever the need arose. In the eighteenth century you still had to make sure you weren't in the wrong place at the wrong time. The residents of Edinburgh would call '*gardy-loo*' every morning before tossing the contents of their chamber pots onto the street from five storeys up. As a respectable passer-by you had to understand that what they meant was *gardez l'eau* (watch out for the water) and call back '*haud yer han*' while making yourself scarce before your wig got covered in shit.

But even without gravity you can run into problems. Weightless shit floats through a spaceship just as easily as the space traveller himself does. On short journeys the first astronauts did what many insects do. These insects collect their faeces in their intestines until they moult, when the shit is shed with the skin. The astronauts ate mostly low-fibre food before take-off and held their shit until they were released from their spacesuits. For emergencies, NASA supplied gloves and plastic bags. It was a messy business, but the bags were designed to adhere to the astronaut bottoms with self-adhesive strips so they could be filled before the shit escaped and soiled the buttocks, trousers and cabin. Today the excrement is sucked out of the anus by means of an ingenious vacuum pump.

On earth, gravity has traditionally been given a helping hand with a splash of water from a cistern. This method has been known since 1596, when the poet John Harington, godson of Queen Elizabeth I, described it in detail in *The Metamorphosis of Ajax*. With its cistern, flush pipe and odour valve, this toilet (known as a jakes) looked remarkably modern. Sir John built one for himself in Kelston (near Bath) and one for his godmother at Richmond Palace. But his brilliant idea failed to catch on. He had forgotten to invent a sewer to go with it, and without good drainage a water-flushing system makes little sense. In addition, the lower classes didn't have the money for such toilets and the upper classes had no need for them. Well-to-do citizens preferred a French invention: the back stairs. Servants could discreetly carry off the turds of their master by going down the back stairs without anyone else in the house noticing.

By the end of the nineteenth century the water closet was complete. The only thing that would change was the fashion.

It would be centuries before the water closet became commonplace.

The name took hold in the nineteenth century as a counterbalance to the earth closet, which was invented by the Reverend Henry Moule in 1860 and which produced compost. You sat on a toilet seat, shat into a bucket, and pulled a handle, thereby covering the poo with soil or ash. But what good was a bucketful of night soil in a metropolis like London? That's when the new sewer system came in handy, especially now that inventors like Thomas Crapper had introduced improvements to the water closet such as an automatic cistern and an S-bend. After that, one *nouveauté* followed another. In 1863 Crapper took out a patent on the self-rising toilet seat, which would lift up after use with the help of weights. Production was suspended after complaints came in from clumsy users who kept getting slapped on the bottom. The deep flusher was more successful. Here your excrement was immediately plunged into the water with a splash, after which it sank into a bottomless pit. This has been the standard model in many countries for more than a century, but in countries like the Netherlands many health services prefer the plateau toilet, in which the shit remains displayed on a plateau until you flush so you have a chance to say goodbye. The deep flusher qualifies as the most hygienic, but it can cost you your life. Flushing away your turd before you can inspect it is just as foolish as emptying your plate with your eyes closed. You should never let reports from your own insides go unread. They might contain a cry of distress: a strange shape, a trace of blood. If automobile seatbelts are compulsory, then plateau toilets should be, too, and for the same reason.

For all the technical innovations, the main purpose of the toilet has always been the same: seclusion. Whether you dive into the bushes at the edge of the village or pamper your buttocks on a heated seat, privacy comes first and foremost. Even a nudist likes to sit by himself behind a sand dune. In order to survive as an individual in a social group you need your own little domain. A territory. Within the cacophony of a colony of seagulls, every gull has its own nest; in the tangle of skyscrapers, every

New Yorker is able to find his own apartment. Your own front door, your own garden fence and your own nameplate all protect you from your fellow humans. Both in the zoo and elsewhere, fences don't so much keep the residents in as keep the visitors out. Cheetahs and antelopes can easily jump over their railings, but they'd rather sit safely behind them. Set your parakeet free to fly around the living room and it will fly back into its cage as soon as it smells danger.

For many animals, having their own territory is just as important as having their own nose, their own stomach and their own intestines. Just as their body consists of organs, their territory is divided into sectors, each with its own function. The same goes for humans. The bedroom is where your bed is, the hall is where you hang up your hat, and the living room is where the biggest television is located. And there's sure to be a fixed place for the toilet. You don't shit in bed and you don't eat in the loo.

Nothing beats your own toilet for the best defecating experience. When you leave the house you just have to make do. Many people hold it until they get back home. So you're very privileged if you have your own toilet at work. But possession makes you vulnerable, as we learn from *Über das Klo* by journalist Horst Vetten:

> I had a *Büro mit Klo* [office with toilet]. Naturally everyone knew that it was meant for the boss, extra-territorial, off limits, *privato*. No one touched my toilet.
>
> After a while I had a falling out with my partner on the staff floor. He did not have a toilet. But one day he used *my* Klo. Our relations reached the stage of a Sicilian vendetta. From his own office twenty metres away he wrote me letters that travelled by roundabout route via the city post office because he wanted to send them registered. And he used my Klo. He would walk past me, bound for the toilet, grinning broadly. I heard him shut the door, then nothing for a very long time, followed by a flush, seat banging, door opening and closing. Him walking past my door again, with the broad grin—I thought I was going to explode.

My career as staff member foundered on this toilet. He stole it from me. He occupied my intimate domain. He emasculated me! I sacked myself—but it was his fault. He didn't give a shit about me!

Vetten's story is one that many animals would understand. Although animals don't have toilets with flush mechanisms like we do, they set great store in having a territory of their own with their own shitting corner. Even pigs instinctively seek out a fixed location for their excrement, as far from their sleeping area as possible. Flies use our lamps as toilets. My mother couldn't stand it, all those little turds on the cloth lampshade over the table. She didn't understand. Why there, of all places, in plain view? At first we thought they were drawn to the light. Later we found that the flies liked the lamps even when they were switched off. It would take years before I learned at biology school what it is that attracts flies. They do it for the company. And to mate. Flies like to be where other flies are, and a lamp is often their first choice because it hangs in the middle of the room and makes a perfect point of orientation for excursions and round trips. And a perfect toilet. In addition, flies are capable of performing a trick that we have yet to master: they crap in two colours. Among the black turds you can clearly distinguish little white ones. The black turds are the real ones; the white ones indicate that the fly has been too greedy and has thrown up. My mother would turn red with exasperation; my father quoted the only German poem he knew, something he had found once in a toilet:

Scheisse in der Lampenschale
Gibt gedämpftes Licht im Saale

Shit on the lampshade at night
Gives the room a mellow light

The flies don't care if the light is *gedämpfte*—mellow—or not. As for other animals, their shitting territory is usually clearly marked by smell. The same is true with people. For a long time after you've used it, your smell keeps other people from taking over the territory you've leased. But what we want more than anything else is to be screened off visually, with a bush if need be but preferably with a real, honest-to-God cubicle. Better a stinking barrel in a cubicle than a high-tech super pot out in the open. My loo is my castle. But there's something quite remarkable about such a castle. Consider this:

Go to a toilet. It doesn't matter where or which: your own, in the pub, at work, or at an elevation of ten thousand metres in a 747. Shut the door and sit on the pot. Look around you. What do you see? That you can't see very far. The mark of a good toilet is that it's small. A cubicle, a cell, a booth. A good toilet gives you the impression that it's been built around the user. He just fits. Custom made. Nicely done. If you build a large house, resist the temptation to equip it with a grand toilet salon. Better to make several small toilets than one big one, so everyone can realise their ideal: your own shit in your own pot. Reaching out from the pot, you should be able to touch most of the walls if the toilet's any good. Not hygienic, but instructive. A biologist in a good toilet can see it right away: these distances are not random. Here Protagoras's statement that 'man is the measure of all things' is all the more valid. Just as an acre traditionally was understood as the size of a piece of farmland large enough to feed one family, so one toilet can be the size of a piece of house large enough for one person to relieve himself in. The architectural dimensions of one toilet correspond nicely with the biological dimensions of a special little area: your individual territory.

Every person has one. It's the area around you within which you will not tolerate any other random human. Anyone who manages to penetrate this space one way or another is either a loved one or an arsehole. Unless there's no choice, like in a crowded train or a lift. There, too, you're quickly overcome by embarrassment. Just exchanging glances can become too much. Try it. Fix your gaze on the train passenger sitting across from you. That's all. No matter what happens, keep looking. First the person being looked at will act as if they don't notice; at the very most they'll cling a bit more tightly to the book they're reading, not a single word of which will now get through to them. Then they'll try to avoid your gaze by looking in another direction. Quickly the avoidance becomes a struggle. Like a snake in the throes of death, the gaze wriggles through the train carriage. But your looking also sets off a series of internal changes in your fellow passenger. Their breathing visibly increases, their heart starts racing, their blood pressure rises, hormones sow panic everywhere, muscles refuse to do what they're told. And all this because of a glance. Your eyes have bored through the invisible wall around your fellow passenger like a battering ram through a mediaeval castle gate.

To keep everyone from going through life with heart palpitations and screaming hormones, we usually respect other people's borders. The reason so many newspapers were once sold at railway stations is not because passengers are excessively interested in the news of the world but because they need a way to withdraw from each other's gaze in crowded commuter trains. We do the same thing today with our mobile phones. This isn't possible in a lift, where there's neither the time nor the space for reading. Here, all eyes are directed past the other passengers' heads and are focused on a counter, which officially is meant to indicate the number of the floor. Breathlessly the lift passengers follow the numbers as they slip by, looking at them with wide eyes as if they were football scores. Once outside the lift, the individual territories expand like air bags to assume their normal proportions. Their size depends on the circumstances. In the city you're lucky if you can manage to keep everyone at arm's length. That's the distance at which you can welcome someone else with a handshake and still hold them back. In the countryside, which is less densely populated, people insist on two arms' lengths. There they shake hands with outstretched arms. The extra distance is compensated for by simply speaking louder, which doesn't really bother anyone.

The individual territory goes with you everywhere like a voluminous, invisible garment. You see it in animals too. Cows like to walk together in a herd. Yet they will never touch each other, unless it's unavoidable or sexual urges are involved. Finches sit neatly spread out along their telephone wire as if they had measured each space. Scientists have measured the spaces for them. Apparently the birds maintain a distance from each other of from eighteen to twenty-five centimetres. If you'd like to divide a telephone wire into equal lengths, you might consider using a flock of finches instead of a ruler. It's a good system. As long as each one respects the others' territory, there are no quarrels. If all your senses decide that your entire individual territory is free of infringement, they give you the 'all clear'. Peace.

If it's peace in abundance you're looking for, go to the toilet. No territory is better respected than this one. To be absolutely safe the lock has to be turned to read OCCUPIED. It seems like such a big word for such a small room, more suitable for war and oppression. In war, on the footpath and in the toilet, the OCCUPIED status alternates with the FREE status. If it's free, that means someone else is free to occupy the country, the footpath or the toilet. At least that's the way occupiers see it. And they cheat, too. Look at all the rulers who have been murdered while defecating. Pope Leo V and the King Henry III of France are only two examples of people who imagined themselves safe on the loo with deadly consequences. But there's not a lot you can do about it. You may be more vulnerable than ever with your trousers around your knees and your bottom bare, but you can't be on guard on the toilet because once you're on guard you can't shit. You don't have to be safe to defecate as long as you imagine you're safe. The smaller the space, the better your chances.

Every biologist is familiar with this illusion of safety from their knowledge of cockroaches. As alien and as remote as cockroaches often seem, that's how familiar they appear to me at other times. The most blissful moments I've ever experienced are those I have in common with the cockroach. I can easily recognise in him that one quality that has kept the cockroach going as a species for 250 million years and has provided me throughout my life with moments of the most supreme comfort: thigmophilia. *Thigma* is touch, and *philia* is to be fond of something. Cockroaches are fond of feeling something on all sides. They want to be surrounded, with their bellies on top of something, their backs underneath something, and their sides pressed up against something. This propensity leads them almost automatically to seek out safe cracks in tree bark and spaces between fallen leaves, behind the wallpaper, or under a carpet. Eels wriggle in the water under tree roots, behind rubbish, into the mud, and within the remains of an old cow carcass. In 2006, while searching for a way to make life in fish farms somewhat more tolerable for them (larger tank? more oxygen?), the researchers from Wellfish discovered that eels prefer 'to lie against something'. People have the same penchant. What's more glorious than being in bed, wrapped up on all sides in down or blankets, only the tip of your nose poking out of the sheets? The cockroach's cracks and crannies are our foxholes, darkrooms, Fiat 500s, sunbeds and the sandpits we dig on the beach. And our toilets. Nice and cosy. Cocooning. The human being is a thigmophile, heart and soul. And stomach and intestines, too, of course.

People are afraid of the big bad world outside. That's why they become social: there's strength in numbers. But once we're together we begin to fear each other. That's all right. Fear is a good counsellor. It keeps you from being overconfident. Truly fearless people are generally never long for this world. Better to take a tip from the animals: the mouse looks around ten times before gobbling a mouthful of bread, the fish is always ready to dart into the waterlilies. A rabbit that doesn't pay attention is an ex-rabbit. If you're an animal you're not safe anywhere. A life like that is hardly worth living, or so you would think. Your nerves alone would kill you in nothing flat. But evolution has found a way to deal with this.

Safety may be a rare commodity, but the sense of safety is everywhere. The rabbit feels safe in its burrow, the mouse nestles under the roof tiles without a care in the world, the eel finds something to snuggle up against. Thus despite all its cruelty, nature is merciful. Even to the human. Our last resort during the day is the toilet. At night there's always bed.

For the human thigmophile bed is the ultimate experience. When night-time falls, when the witching hour strikes, when the cold creeps in and the cockroaches actually abandon their safe cracks, human beings dive under the blankets. Powerless while sleeping, prey to burglars and murderers, they feel safer than ever there, lulled to sleep by the murmuring of touch receptors.

One person swears by a cover of light down, another by the weight of real blankets, but the principle is always the same. The womb-like warmth that the down or blanket radiates does not come from the thing covering you but from yourself. It's you that warms the down or blanket, and not the other way around. The glow comes from the depths of your own body. It isn't the arms of Morpheus but your own intestines that warm your skin, even on the outside; the blankets only serve as out-testines. You've been turned inside out, as it were. That's why after a few minutes the blankets no longer feel like strangers to your body. Man and blanket melt together; bed becomes man, man becomes bed.

In a spaceship there would be little sleeping done if you were to float weightless through your capsule. That's why Wubbo Ockels came up with a sleeping bag for outer space. During his journey, the Dutch astronaut crept between the blankets with an air hose. 'When you pumped up the hose, the sleeping bag pressed tight against your body. It made for very comfortable sleeping.'

Many people wouldn't sleep a wink in such a situation. Here on earth they already break out in a sweat just by entering a lift or a tunnel. This is called claustrophobia. *Claustrum* is Latin for enclosed spaces, *phobia* is Greek for fear. The fear of enclosed spaces has two components:

fear of suffocation and fear of being trapped. Claustrophobia is the opposite of thigmophilia. Yet often these are two souls in the same body. The air-raid shelter you fled to in order to feel safe makes you feel anxious at the same time because you can't get out. Fortunately the dangers during peacetime aren't so threatening. You can venture outdoors and stay there all day, which makes coming back to your cosy lair in the evening all the more delightful.

Heaven for a thigmophile is hell for a claustrophobe. On the toilet, too. Even people with a bit of claustrophobia soon become anxious in such small cubicles. But something has been invented to deal with this. Take a look at an old-fashioned outhouse with its cramped surroundings and your eye falls on a detail as secretive as it is indispensable: the window. There are good toilets in abundance without flushing mechanisms, but a toilet without a window doesn't count. A window seems inconsistent with the need for seclusion in a toilet, and that's exactly what it is—on purpose. It's a safety valve to keep the seclusion from becoming too intense, much like the valve that allows the carefully built-up pressure in a steam locomotive to escape before it gets too high and the boiler explodes. If you used to feel unsafe riding around in a Citroen deux-chevaux, which wasn't such a crazy reaction to being in a biscuit tin on wheels, you'd clap the side window open and breathe freely once again.

In order for you to feel comfortable in a closed space there has to be some contact with the outside world. A standard feature of a bunker is the chink in the wall for scanning the surrounding area (and through which the enemy will finally toss the grenade). The most important part of a submarine isn't the propeller or the tail fin but the periscope. To see without being seen, that's what makes the periscope perspective so attractive. Which is not to say that you can actually see out when you're sitting on the toilet. Usually the window is simply too high for that. But it doesn't matter. It's all about the idea of not being a prisoner. A window is a good addition to a toilet. Unfortunately, in modern building construction it doesn't stop with the addition of a little window. Glass is taking over the entire facade.

The ones who profit most from the glass house rage are the curtain manufacturers. When evening comes, the people who live in glass houses shut their curtains with a sigh of relief. No shutting curtains during the day, when they have to make do with the only space not yet infiltrated by the outside world: the toilet. But you only go there when you really have to go. It's almost impossible to escape the new dictatorship sweeping the world under the banner of transparency. Everything and everyone has to be transparent, from buildings to opinions, from trams to politics. It sounds attractive, transparency—free of shady practices, underhandedness, covert goings on, or guile and deception; what you see is what you get, and what you get is what you see. With transparency you know precisely how things stand. The problem is that most things aren't precise. A lack of privacy turns a transparent building into a closed institution. If you listen carefully to a transparent argument you recognise the thinly veiled flimflam. Keeping your cards close to your chest has always been a good strategy. True thigmophiles are horrified by transparency. They live by the grace of non-transparency in the dark crack in their arses. Their pleasure is epitomised by animals like the caddisfly, whose house is attached to its body, both fused into a single entity. It is the crack in its own arse. Caddisflies live in protective cases made of the rubbish they find in river beds. One species spins twigs together, another does the same with pebbles, and yet another uses minuscule snail houses (tough luck to the snails still living in them).

No matter how remote and odd a caddisfly may seem to the human eye, the feeling of being entirely encompassed, closely fitted, with only head and foremost claws sticking out, is not the least bit odd. We spend more than two-thirds of our lives walking around in clothing that leaves little more than our head and hands exposed. Clothing is a mini-house, the smallest conceivable one-man tent, a mobile crack in the arse.

~

Yet you can also be cramped for space. There's no room in your clothes to install a decent toilet. Now what? Nothing is worse than pooing or peeing in your trousers. In order to allow for the timely release of faeces and urine, openings have been strategically placed in your clothing. The first requirement is that when closed these openings must look as if nothing is going on underneath. Openings in the wrong place or at the wrong time are not tolerated in a society that prides itself in openness. A hole in your sock today is already a scandal.

Whether Scottish men really have no underpants under their kilts ought to remain an everlasting question. At one time Dutch women also had people guessing. Back when the Jordaan district of Amsterdam was still a slum, the women wore split underpants under their long woollen skirts. Standing with their legs wide apart, their skirts as round as hoops, they pissed in the middle of the street with the sanctimonious expression of a child peeing in the swimming pool. Men had both a fly in the front and a flap in the back to allow access without having to drop their trousers. Unfortunately these styles are no longer available. Even underpants don't always come with flies anymore, which means that men, too, have to pull the whole business down. So there you stand with your bare arse, and your most vulnerable bits exposed, of course. Never was the need for a fully sheltered cubicle greater.

All the horrors of the past aside, there are trousers that are expressly intended to be shat in. You yourself wore them for years, and it's quite likely you'll have to wear them again later on for several years more. Many people start and end their lives in nappies. Fortunately this great discomfort is compensated for by the thigmophilial pleasures of the beginning and the end, in the cradle and the grave. Unfortunately, a thigmophile reaches this high point when he is least aware of it. That's all the truer for the thigmum of all thigmums: the womb.

As a fertilised egg grows into a baby, the womb grows along with it. When empty, the womb is an unsightly, flattened sack with the capacity of a shot glass. The embryo stretches the sack into a succulent pear with the neck pointing down, a neatly maintained dwelling in the slums of the belly, behind the bladder full of piss and in front of the shit-filled rectum. There it is nursed by juicy flesh and rocked in amniotic fluid, setting the benchmark for your sense of security. You'll never again be so well protected. For many animals, the party's over as soon as they leave the womb. A foal has to stand on its legs right after birth, a duck immediately starts swimming with its mother. Human babies would never survive something like that. They come out of the birth canal half-baked and are pampered in an artificial womb long before they can be let loose. From the mists of antiquity until well into the eighteenth century, newborn babies were swaddled as a matter of course. Wrapped in cloths and stiff as mummies, little arms and legs encased like flea larvae, babies were consigned to their embryonic paradise for months. This made them more or less equivalent to the other little monkeys, who cling to their mothers' fur for that extra-uterine sense of security. You'd never manage this as a human baby, though; evolution has heartlessly robbed your mother of her fur.

Swaddling has now fallen into disuse. Many people who care for infants swear by freedom of movement for the whole baby, although no one has ever demonstrated that Mozart would have composed better or Napoleon would have slaughtered more people if these men had never been swaddled. All we know for certain is that babies in countries where swaddling is still practised cry less and sleep more than the Western breed, who thrash about freely in their cradles and are pushed around in baby carriages. What it's all about, of course, is striking a golden mean between freedom and security. For such a mean to be golden, however, children need a good helping of security in their youth in order to take on the world as adults later on. But even with a happy childhood behind him a person will have to soak up the security of beds, toilets and telephone booths from time to time before surrendering to the dream destination of every thigmophile: the coffin. By that time there's little left to enjoy, but until then it's all preliminary fun.

When you reach the end of the line, the need for freedom of movement gives way entirely to the need for security. It isn't until your relatives have packed you in snugly that they dare commit you to the earth. In order to convince themselves that you'll like it there, the coffin is usually furnished like a bed. But if there's one thing you are not free to conclude it's that you are what you are: dead. You're laid down as if you were asleep. It would be more convincing to turn you on your side, your legs drawn up, preferably in the securest of all secure positions, the foetal position. Your eyes and mouth shut, your relatives entrust you to the bosom of Mother Earth. But she isn't keen on eternal rest. As far as nature and the funeral director are concerned, your decomposition couldn't happen fast enough. Surrounded by the silence of the cemetery, the body processing machine is running at full tilt. If you could hear the gnawing of insect mandibles, the sucking of trillions of bacteria, and the hissing of all the chemical reactions, the cemetery would thunder like an industrial zone. While coffins were once made of stone in order to keep the riffraff out, today the materials are governed by strict rules so the riffraff can be welcomed with open arms. The idea is that within a year only a few of your bones will remain, and whatever is cleared away after ten years is hardly worth the trouble.

A coffin is a phone booth without a phone, a canopy without a bed, a cubicle without a peepshow, a hut whose only company is death. And a toilet without a toilet bowl. That no longer seems necessary. But what's left of you after a year, except for a few bones? After your death you perform your last feat of strength: you poo and pee yourself away. Even though you don't eat anymore after you die, your metabolism plods on undaunted—but with yourself as its food. A corpse eats itself up. Without teeth. There's little left for the worms. Long before any worm strays from the upper reaches of the earth, finds the coffin, and bores into it, the body has fully consumed itself, assisted by its bacteria. All your life, your bacteria have been waiting for this chance, but they were always beaten off by your immune system. As soon as you die they make their way through your organs, pillaging and plundering. Some of them can't even wait until you're completely dead. If the resistance of a terminal patient is sufficiently weakened, the bacteria tear down the barriers that have been standing in their way all this time. This begins in the intestines, where most of the bacteria live. The wall that once held them back is now penetrated as they make their way to your liver, and you're eaten alive the way a vanquished zebra is eaten by a lion. To steal a march on all those attackers, the body cells then eat themselves up. They're full of suicide capsules, the lysosomes, which are filled with enzymes whose job it is to clear away cells during the continuous construction and demolition activities that keep the living body in shape. The decomposition products don't necessarily leave the body through the intestines (or what is left of them); incontinence has now taken control and the whole coffin is smeared with bodily waste.

To prevent the coffin from becoming soiled during the funeral, a fresh corpse undergoes careful preparation. Intestines and bladder are emptied, after which the throat and anus are kept in check with a wad of cotton. The risk of undesirable smells is already considerable. If the corpse can just manage to delay its decomposition until after the funeral, it won't need a toilet in the coffin. The body can stop trying to keep up appearances and will be allowed to blend with its own sludge.

But what about back in the beginning, in the womb? No sanitary facilities there either. Going nine months without plumbing seems like a tall order. Is that why the maternal belly swells to such vast proportions? No. The waste products of the unborn baby are sent back via the same route by which its food is delivered. In the placenta, baby blood and mother blood swap food and waste. And that's a good thing, because at the end of the pregnancy a baby can drink three-quarters of a litre of amniotic fluid a day and pee out the same amount. While the mother lovingly hauls her belly around with a look in her eyes that you only see among true believers, the darling baby uses its mother as a piss-pot. And Mum pulls the chain.

A baby doesn't defecate until after its birth. There's too little to eat in the mother's belly to create anything worth shitting. When the baby is born, its stomach is barely recognisable, an empty little sack hidden behind the inactive intestines. At fifty grams, the baby's entire intestinal tract weighs less than a tenth of that of an adult. Shit only accumulates there in the latter stage. The baby has ingested all sorts of things along with the amniotic fluid: exfoliated cells from the skin, the oesophagus and the trachea; lanugo, baby skin grease, glandular fluid, blood, urea, and enough bile pigment to colour the intestinal ooze green. By the end of the pregnancy the whole large intestine is full of it. Sometimes the baby lets a little of it run into the amniotic fluid, but a good child waits until shortly after birth. This syrupy, dark green-brown stuff is properly known as meconium and is easy to distinguish from the yellowish fledgling excrement that begins flowing so abundantly after a few days.

As baby or corpse you don't have anything to worry about. In your cradle or coffin you're safely sequestered from the rest of the world. The hardest bit is life between cradle and coffin. You spend a long time living in the public realm, hoping that the borders of your individual territory are respected. But no one can see those borders. They're built on something ethereal, like a sense of unease. The only way you can tell that they've been transgressed is if you feel uncomfortable. And you wouldn't feel entirely at ease in public unless actual walls were erected around you. That isn't so hard to imagine. Take that man out there in the square, encapsulate him in a steel-and-concrete construction following the lines of his borders, add a few windows so he can see out, and the result will look vaguely familiar: a phone booth. It's partly because you don't see phone booths very often anymore that every time one of them comes into view I hope there's someone I need to call.

The telephone itself hangs in its booth like bait in a trap cage, the perfect excuse for disconnecting yourself from humanity and, without any embarrassment, bringing yourself to express the most personal intimacies in the midst of the repudiating masses, like the singer Jim Reeves in his song 'Put Your Sweet Lips a Little Closer to the Phone'. Try coughing up lyrics like that on a mobile phone as you stride down the high street. Besides the lack of seclusion, a mobile caller has to do without the immobility, be it voluntary or involuntary. While the rest of the world spins around you, you remain the unmoved centre of your universe there in your little structure. Compared with a 'public talking facility housed in a booth', as the phone company calls it, these walkie-talkies are just cheap baubles. With a mobile phone you don't escape the masses; you get sucked into them. I look at this and shake my head. If you're able to accumulate your pee like a decent person until you reach the next available toilet, why can't you hold your yacking until the next available phone comes along? Mobile telephoning is a verbal form of incontinence.

Long before the phone booth began vanishing from the street scene, the thigmophile had been driven into the pub. Here, until this century, there was a special place of which I preserve many warm memories, between the door marked L and the one marked G: the door marked T. As soon as you opened it you were assaulted by a cloud of cigarette smoke, stale beer, coffee grounds and drenched all-purpose carpeting that I otherwise knew only from student rooms and regional newspaper editorial offices. The full ashtray was balanced on a stack of telephone books, incomplete—many of the missing phone numbers were scratched onto the wall. The fact that you could barely move in this hut, foot on an empty crate, back against the stairs under which the T was usually to be found, didn't bother me in the least. The only difficulty was finding the telephone amidst all the clutter. When you finally got your hands on it you reported that unfortunately you had to work late. At no other time and in no other place has the phone company so fully cooperated in the spreading of lies as then and there. The entire booth smelled provocatively of sin.

What was striking about the pub phone booth was its correspondence with the adjoining toilets. The same form followed the same function. In both the T, the L and the G, one eliminated what one had to get rid of. That brought with it certain miscalculations. I recall a lively evening in the Amsterdam pub known as the Karpershoek, where the G, the T and the L were still neatly lined up. At a certain point we saw a yellow stream trickle out from underneath the door marked T and into the pub, soon followed by a relieved Uncle Theo, who staggered back to the bar to fill up his bladder once again. Something like that seems out of the question today, in the age of the mobile phone. No one pisses in the phone booth anymore; one makes one's phone calls on the toilet. Pub patrons go outside to politely make their calls so as not to disturb anyone. But apparently there's no fixed location for such activity out there. Recently I saw someone making his call against a lamppost.

Armed with their mobile phones, people have demanded the right to spout their drivel at the slightest pretext. With the help of satellites, earth's entire atmosphere has been filled with human babble and chatter. And woe to the provider who has neglected to arrange for reception in the middle of nowhere. All the more surprising is the lack of access for that other form of drivel. If you suddenly find you have to pee, you're better off at home. Apart from exemplary cities like New York and Tokyo, suitable peeing facilities along the public thoroughfares are few and far between. The ones that were there are disappearing from the street scene with the speed of the telephone booths. For people with small bladders, it's getting increasingly harder to plan a walking tour through the city. Personally I can't escape the impression that in places like Amsterdam you see more and more people walking around with anxious looks on their faces, but that may be for other reasons, of course.

The disappearance of public toilets is something to mourn, not only from a physiological point of view but also from the perspective of urban beauty. In the past, especially during the Victorian era, real peeing palaces were erected—cheerfully styled, elaborately decorated, often patterned on something else, like a castle, as if to provide a festive touch to the minor physical pleasure being offered there. In London there were also underground piss palaces called Halting Stations. When the first one was opened in 1855, Josiah Feable expressed the newest step on the road to progress in the presence of Queen Victoria:

> Down gleaming walls of porcelain flows the sluice
> That out of sight decants the kidney juice.

Thanks to the opening in the front you could wipe your backside in a Roman toilet without having to stand up. Not with paper but with a sponge on a stick.

Not much later, the celebrated pissoirs appeared in Paris under the name *vespasiennes*. This was in honour of the Roman emperor Vespasian, who was said to have provided public toilets back in the days of antiquity. The remains of the Roman facilities can still be seen. There was room there for twenty-five men to sit side-by-side on stone latrines, shitting and chatting together. We get the impression that these were people as tidy as they were convivial, with outstanding sanitary facilities, but according to the Limburg archaeologist Gemma Jansen nothing could be further from the truth:

> From finds we learn that conduits, gutters, and drainpipes were often blocked, and that the floors were therefore flooded with water and urine. Sometimes toilets and baths were half caved in but still in use. In short, it was filthy and it stank.

This could be a description of contemporary Amsterdam. Yet a century ago a kind of toilet architecture was being flaunted there that was no less impressive than that of London and Paris. On the Nieuwezijds Voorburgwal there's a sweet little pine house in chalet style that still bears witness to those glory days. It's a confection of a building. You'd never believe by looking at it that something so charming could ever have been built for something as trivial as piss. Now it's a little restaurant, where tourists devoid of any historical sense come to partake of fashionable titbits. Seldom have we been so harshly confronted by the victory of eating over peeing. It's a clever chap who, at the moment of greatest urgency, manages to find the nearest urinal on the Singel in plenty of time.

But it still exists, the unsurpassed apex of Amsterdam street architecture: the *krul*, or the curl. Like an ornamental ribbon of sheet iron shaped by a giant florist with giant shears, this urinal folds itself around the grateful user. What a pleasure it is to take a breather there while peering through the ingenious pattern of little holes in the enclosure and watching the seagulls and people along the canal. Rarely do I pass one up when the opportunity presents itself, even when there's a whole row of them—a veritable stairway of sluices—as there is on the Singel, each one laying claim to the very last drops. And what a glaring absence it is when one of them is out of service or, worse yet, has disappeared! When that happens I feel like my cat when he's startled by the vacuum cleaner and takes his standard escape route through the house, only to discover that there's a side table missing which causes him to miscalculate his jump—one of the few times I think I've ever heard a cat swear.

You can see from the elegant lines of the *krul* that it was designed by a genuine architect. That was Johan van der Mey (1878–1949), who perfected it in 1915 and whose luxurious design for the Shipping House (*Scheepvaarthuis*) in 1928 makes him a pioneer of the Amsterdam School. But there are also later structures that exert an attraction on the type of men whose interest in ornamentation is unusually acute. In the urinal across from the Hortus Botanicus, a pink lamp testified for years to the pleasure the city's gay community took in helpfully changing the light bulbs. Even more popular during the emergence of Amsterdam as the 'Gay Capital' was the pissing palace below the Munttoren. The city declared it was being improperly used, which partly accounts for the disappearance of many other urinals from the streets. They've been replaced in the city's entertainment districts by plastic pissing towers, where four men have to struggle to avert their eyes from what's taking place right under their noses on either side of them. Once again, an innocent little pleasure has deteriorated into distasteful spattering.

The *krullen* unite elegance with the greatest possible technical simplicity. Essentially you're just peeing against a wall, albeit a wall made of a special kind of stone that can take a lot of abuse. When it comes to cleaning them, the city relies on the theory that urine pollutes and cleanses at the same time. Mildly acidic, and entirely sterile when it leaves the body, urine makes an outstanding cleaning agent. The bacteria that give urine its characteristic smell don't start working until they're outside the body, and when they do their diligence is not universally appreciated. So from time to time the sanitation department comes by to spray the *krullen* with water, although no one believes they've ever actually been inside them. Women tend to walk a wide arc around each *krul*, but for some men the smell of piss exerts a certain attraction, like a much-used lamppost has for dogs.

The toilet facilities on public transport are a case apart. During the age of the steam engine none of the trains had any facilities at all. The trains did make regular stops, however. The locomotive had to fill its water tank and the passengers had the chance to empty theirs. Today the Dutch trains stop too briefly at each station to allow passengers to relieve themselves. Many stations don't even have toilets. Fortunately, most trains have toilets on board, but even these are being nibbled away. In around 2010 the Dutch railways acquired a whole fleet of trains that were entirely without toilets. For what reason? By eliminating toilets you could survey the inside of the whole carriage from front to back, a boost for 'security' and—oh, yes!—for 'transparency'.

4

Shit Central

How do you make a lion from a lamb? A good question for a zoo director. Lions are more popular with the public than lambs. Fortunately the solution is simple: put the lamb near the lion. In no time at all the lamb will have become a lion. That's how you make cats from mice, mice from cheese, cheese from cows, and cows from grass. Carrot becomes rabbit, rabbit becomes human being. In the beginning you're no more than a single fertilised egg. The rest of you consists of fish, peanut butter, meatballs, pork loin and potato salad. A cow's ear becomes a human toe, a lamb chop becomes a human ear, the nose of a salmon becomes the pimple on your bum. The brains I used to write this book were once located in a pig's backside, the fingers I used to type it consist partly of rabbit paws. That's pretty clever, such a conversion, but it's also scary. Imagine them getting stuck halfway. The fact that what you are now was first flounder and pig is still frightening. Molecule for molecule, an animal becomes a human. It can't get any more intimate than that. Too close for comfort. That accounts for the flourishing business in organic steak and unsprayed hot dogs. Many people are becoming vegetarians. From now on their brains will come from turnip greens, their fingers from asparagus, which only makes the miracle greater. Have we found the philosopher's stone? We still can't extract gold from iron, but every day we make human beings and *joie de vivre* from chicken and apple sauce. No matter how clever it was of Zeus to turn into a swan, and for Jesus Christ to become incarnate, for an ordinary mortal metamorphosis is child's play. Literally. To become an adult, a child has only to eat himself up. Innocent children's tissue is transformed into a tremendous penis or massive breasts, brains full of innocence into a tub full of double entendre. Caterpillars eat themselves up in order to pupate into butterflies, amphibians eat themselves from fish to frog.

All that a living being needs, human or otherwise, in order to incorporate another type of creature into itself is a magic vessel: the gut. Within the intestines of all the animals in the woods and all the dwellers in the cities, plants and animals are constantly changing their identity, day in and day out. Living nature seems to resemble a box of Lego or a Meccano set. First the food is broken down into its building blocks, then the building blocks are transformed into a new creature. A very complicated business. It takes two days alone just to demolish the food. It all bears a suspiciously close resemblance to glass recycling, that labyrinthine process of crushing bottles to shards so that new bottles or maybe jam jars can be made. What a waste of energy.

It takes a lot of energy to make a bit of a human from a mouthful of food. Yet food provides energy. Most of it is not used for building but for heating, and that's good. A human being eats a tonne of food a year. If you were to convert all those kilos into kilos of human being, you'd end up as heavy as a whale. In order to maintain your weight, it's a good idea to burn up most of the food in due course. Fortunately there's a big demand for fuel to keep your body going. Your whole body is alive with sputtering and bubbling; nerves flash, arteries beat, turds are eager to find an escape route. Every animal and plant is sizzling: it gurgles and hisses, hormones go screeching, juices are absorbed, colours and smells are dispersed. All of nature is convulsing from the living factories in the industrial zones borne by millions of tiny legs; factory smoke wafts from billions of anuses, big and small. In each of those countless organisms, substances are transformed into something else until they drop from exhaustion. What are they making there, anyway? And why are they working so urgently? The end product of all that effort, all that fuss and bother, is life. And life goes on. And on. And on. And on. For three billion years now.

The entire surface of our planet is seething, thanks to life. Sometimes it even threatens to boil over. The engine behind this global commotion is the sun. Every day the sun drenches the natural world in hundreds of times more energy than all our power stations put together are able to extract from coal and oil, and life gets a tiny share of that solar energy. Essentially, life is no more than a detour along which the energy from a glimmer of sunlight is able to play outdoors before being swallowed up in heat or disappearing into the depths of the universe. Crowing with pleasure, solar energy gurgles on through tree, human and worm. Animals, plants, people: we're all the sun's playthings. It's a lively chaos for those who see the fun of it, and a vale of tears for those who can't bear the idea of finitude.

Why are we here on earth? We are on earth to metabolise. Metabolism provides the energy that keeps life going. In the words of biologist D'Arcy Thompson:

> Energy is life, and life's currency. It unites and divides all living things; its flow from one place to another controls everything from cells to forests.

Long before the concept of 'energy' acquired its modern connotation in the nineteenth century, it was generally thought that something like it flowed through every living organism. Taoists called it 'chi', Hindus 'prana'. Until the twentieth century there were also Christian biologists who believed that life could not be understood merely in terms of scientific formulas. There was also said to be evidence of a life force or *vis vitalis*, whether sent by God or not. But when it became possible to determine with greater and greater precision that a plant or animal uses up exactly as much energy as it takes in, these earlier voices were forced to admit defeat. Life, too, complies with the pure physical law stating that energy always remains constant. At the very most it changes form.

The art of living is all about nibbling as much as you can from the available energy before you die. Plants capture as much solar energy as they can with the help of leaves that are as plentiful, as big and as close to the sun as they can be. Humans churn the energy out of their food in their intestines, a place that not a beam of sunlight ever penetrates. They leave the capturing of sunlight to the green plants. That seems like an awkward arrangement. Why would you content yourself with second-hand solar energy? Why don't we capture the sunlight ourselves? Why didn't evolution give us green leaves so we could tap the sunlight directly? The answer can be found in every garden. There they are: the roses, the hydrangeas and the lilies-of-the-valley. They live from the air and the light. Maybe that makes them happy. But all they do is stand there. They can never check out the neighbour's garden to see if it's nicer, they can never search for a sunnier corner on their own initiative. Plants don't have legs, true, but even if they did they wouldn't get very far. They simply don't have the get up and go to move themselves forward. With their leaves, plants can grow and blossom, they can attract insects and feed people, but even the biggest giant sequoia has to spend all its centuries in one and the same place. Plants lives their lives as plants.

If you want to have enough energy to live an active life, then your only choice is to turn to crime. Instead of laboriously extracting energy yourself from the sunlight, you rob the hard-working plants of the solar energy they've stored in their tissues. With just a few mouthfuls you get the harvest of energy that took months to glean. See how energetically the antelope leaps through the immobile grass, see the birds fly on the energy they derive from seeds under house arrest, hear the rutting buck bellowing from the strength that's been churned up from static tree bark. But even their lives are not carefree. They have to be on guard against the super bad guys who see such plant plunderers as even more densely concentrated energy packets. With one neck bite, one gash of the beak, a lion or eagle handily relieves an antelope or hare of all its diligent labour; with one visit to the butcher, a human being takes home the result of a lifetime of chewing and re-chewing. The advantage of eating meat is that you need less of it: you get more energy from one chop than from a whole bale of hay.

You eat to live. But living still doesn't use up all the energy you get from food. There's plenty left over for a hefty pile of shit. The amount of energy you lose with your faeces is not precisely known. 'In man there have been few investigations in which the excreta are collected and their heats of combustion determined,' writes Scottish physiologist Kenneth Blaxter, not without understatement, 'no doubt because such tasks are not pleasant ones.' It has been estimated that 6 per cent of the energy in your food ends up in your shit, 3.5 per cent in your pee, and 0.5 per cent in your farts; all told that's one-tenth of your energy budget.

If energy is the currency, we're talking about 10 per cent of your life. The idea that you actually lose some of your vitality with your excretions was generally accepted for centuries. From the ancient Greeks to Christian folk religion, philosophers and healers believed that everything that leaves your body—sperm, menstrual blood, sweat, tears, breath and even your voice—is animated by the life force in your innermost being. Shit and urine were considered the most life-filled, since they briefly retained their warmth. Thanks to their vitality you could use them to make all sorts of medicines or even work magic with them. Something of this conviction lives on in the custom of criminals defecating at the site of their crime. With the help of its *vis vitalis*, the turd stays on the lookout while its owner makes himself scarce. There's no time to lose, however, since a turd only works as long as it's warm. A cold turd loses all its strength. So the crook, with his belief in the old superstitions, is properly observing the main laws of thermodynamics, which hold that all energy is ultimately reduced to heat.

Unfortunately for crooks and mystics, the *vis vitalis* does not exist. Even the best burglar or the most isolated hermit runs on the energy provided by chemical processes. Today you can buy all the necessary chemicals at your local supermarket. In every can of beans, every pork chop, every container of cottage cheese there's enough power to cause an explosion. Energy galore. But how do extract it? Put the meat and the beans on a plate, pour the cottage cheese on top, and nothing happens. Not a whisper of a chemical reaction. It's as if there were a brake somewhere. In order to get your food to break down chemically and thereby release its energy, you first have to supply a little energy. You have to lift the process over a threshold, as it were. Start it up. Give it a crank. To light a campfire you first have to make fire with a match. Only then will the twigs burn and the energy be released in the form of heat. It's not all that different in your body. Beans, pork chops, and cottage cheese are actually burnt up in order to release their energy. It sounds risky. Fire has a way of quickly getting out of control. How do you burn meat in a body made of meat? If you're a woodworm, how do you burn your wood without setting the whole antique on fire?

Good questions. But first you have to get the whole process going. How do you start a bean? Hold a match to it and you're left with a lump of coal and a burnt finger. Put it in a nine-metre glass tube with the rest of the beans, the pork chop and the cottage cheese, and nothing happens—unless there's a laboratory attached to the tube that will unleash a whole series of chemical processes. Stick the whole business in a living intestine seven metres long and enough energy will be produced for you to walk around all day, to laugh and to love. The secret of the intestine can be found in these extraordinary elixirs, the enzymes. A drop of this and two drops of that are enough to get the chemical processes going. Fortunately, there's no fire involved. For combustion to take place, oxygen is necessary by definition, provided by means of respiration. But the process works just as well without fire as with it—or better even—provided that the combustion entails a large number of steps, each with specific catalysts that help each sub-reaction over its particular threshold. For every kind of food there's a separate enzyme. These are the juices with names like lipase, amylase, and peptidase, which have robbed many a student of the desire to continue their study of biology. Lots of people know what goes in and what comes out, but what makes the sun shine again in the darkness is a mystery to them. Their intestines are the longest black box in their body. Yet you can easily get a glimpse of how the intestines work—at the table. Choose a dessert that has lots of kiwi fruit in it. The digestive process begins before you take your first bite. The enzyme in a kiwifruit is the same as that in your stomach, an enzyme that is simply wild about animal products like dairy or gelatine. Kiwifruit loves ice cream and gelatine desserts more than anything else. If you don't move fast, the kiwifruit will eat the dessert into snot before you can even get started.

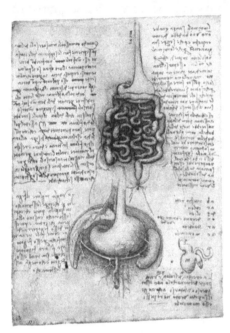

Leonardo da Vinci (1452–1519): The digestion machine.

It's striking how little interest people show in the forces that keep them going. They have a vague notion about the combustion of food in the belly, but most of them let it go at that. Biology is something outside, not inside. The more distant it is, the more we know about it. Exotic animals fill the television screen, exotic palm trees entice us to exotic beaches. We have clear images in our minds of the Ganges, with its praying bathers and bathing prayers, burning corpses and sky full of vultures, but the intestinal river that winds through our bodies is something we rarely try to imagine. More expeditions down the Nile and the Niger have been broadcast than those down the small and large intestine. Towards the end of the era of the great scientific expeditions, the American writer George S. Chappell (1877–1946) decided to take his own voyage to the human interior. In *Through the Alimentary Canal with Gun and Camera* (1930) he describes his adventures in search of the sources of the Bile. The other three members of the expedition prefer to remain anonymous ('one never feels quite the same toward a person who has looked one's liver squarely in the eye'), but they show up promptly for the adventure ('Be here at six-thirty sharp. And don't forget your rubber boots').

> Before us lay who knew what terrors of attack! what onslaughts of infuriated bacilli, scissor-jawed microbes, fierce phagocytes and, most dreaded of all, wild-eyed heebie-jeebies which give nor take no quarter.

Travelling in a collapsible canoe called the *Rubber Duck*, the expedition crew enter the locks of the Oesophagus and descend to the intestines. Danger lurks at every turn:

> Several times we seemed to be bearing down on wicked reefs or about to be dashed against the living wall, but a quick turn of the wheel swept us by in safety. One of the gravest elements of danger lay in the traffic, for as usual in this part of the trip the way was clogged with many heavy-laden food boluses southward bound. These clumsy craft took up most of the room. They are too heavy

to manoeuvre for position, and we had to get by them as best we could.

Along the banks of the Upper Bile, Chappell hears a faint, curiously rhythmic sound:

> *Haemoglobins!*
> My hair rose as I thought of being enveloped by one of these ghastly wet-blankets. A headline in the Livermore *Leader* flashed through my mind, 'Explorer Found in Bile Valley. Asphyxiation Indicated.' I saw my body being discovered...the subsequent obsequies... weeping relatives...you know how it is.

This cannot end well. Strikes break out in Gastritis ('they're all reds'), colic spreads, and there are head winds all the way up to Colon-by-the-Sea. The crew members are just able to purchase postcards of the sights in the south that they hadn't seen yet—the Appendix from Caecum, Caecum from the Appendix, the Topless Towers of Ileum, the Everglades of the Upper Colon, the Elks Club at Colon-sur-Mer—before a coppery cloud eats up the landscape. The Bile has overflowed its banks! With a powerful eruption from the interior the entire expedition is washed to the north on a tidal wave, through the lock at the Gizzard, and back into the outside world.

Seventy-five years later, this pastiche of an adventure novel met its medical match on Dutch TV. This time the visitor was obviously quite capable of fitting into someone else's intestines, since he was a gnome. Prikkeprak the Gnome (played by Arjan Ederveen) works for the Academic Hospital in Utrecht, where Professor Tineke Poortvliet (Tosca Niterink) has developed a revolutionary method for removing intestinal polyps. As the hairy buttocks of the patient, Mr Struikebos, come into view, the professor explains:

> Prikkeprak the Gnome is at the entrance and is waiting to begin. In just a moment we're going to insert him into the rectum. The gnome is equipped with an oxygen tank. He's carrying a video

camera so we can watch the entire procedure, and he has a field telephone so we can communicate with each other.

All right, there he goes. First he's going to examine the anus a bit. Can you feel that? Now he's going to stick his little hand in. And now he's going to try to get his cap in as well. Shouldn't you put a little vaseline on that cap? He's moistening it with some spit.

With the tip of his cap leading the way like an awl, the gnome disappears into the hindmost part of the intestines that were closed off to Chappell. On the monitor we see Prikkeprak get to work deep underground. When he finally finds the polyp ('Mr Struikebos, are you sure you haven't eaten anything? It's really hard to see in here!'), the gnome takes a little hatchet and starts chopping away with the enthusiasm of a happy worker, accompanied by suitable songs ('Hi-ho! Hi-ho! Hi-ho! Lalalalalala'). This takes quite a while, because 'this thing's a real brute'. His joy is all the greater when the job is done.

Prikkeprak the Gnome at the entrance to the rectum.

Intestinal tourism is no longer popular. Today the human adventurer has been replaced by a single eye in the form of a camera on a string: an endoscope. With the string, actually a flexible tube, you can inspect almost the entire intestine from within. Little robot hands come back out with the souvenirs they've harvested. This makes it possible to examine any growths for malignancy. But even an eye with hands can involve inconvenience. First the patient must empty his intestines. If he doesn't, the eye won't be able to see anything and the little hands won't know what to grab. Here a laxative is as effective as it is disgusting. Fortunately, as an amateur you don't have to go inside to get an impression of the state of your entrails. You can also wait until something comes out. Call it the volcano method. As a volcanologist you're wise not to descend straight into the sulphurous hell of the crater. It's best to wait until the volcano has erupted spontaneously so you can examine the innermost recesses of the earth somewhere on the outside at your leisure.

A cautious gastroenterologist will wait until a person has to vomit before studying the stomach. In one brief but powerful eruption the person's abdominal contents come to meet the light of day. It's rather startling. What until only recently had smelled and looked so appetising on the plate has now become revolting mush—not shit by any means, but just as nasty. With the exception of gastroenterologists, very few people are fascinated by vomit. An anatomical pathologist might be intrigued, but then there would have to be an interesting murder to investigate. I myself once tried to make fake vomit. On TV I was teaching children how to feign illness in order to avoid having to go to school. Following some tips on whipping up imitation blood, forming really ghastly sores, and pretending to chop off your fingers, came the highlight of the program: fake puke. I mixed cold tea with garlic, vinegar, food scraps, cinnamon, coffee and beetroot. It took a long time. Finally I called out, 'There! Your puke is ready!' But the director wasn't going to let me get away with that. She wondered if I'd be willing to take a big swig. In a burst of obedience

I complied with her request. Even before the fake puke could reach my stomach a wave of real vomit came up, so that fake and real together were sprayed all over the camera lens. The director was satisfied. I less so. The cameraman least of all.

The contents of the stomach are easy to examine, if necessary.

It's simple. Once the vomit centre in the brainstem has decided it's time to puke, an automatic response is triggered that cannot be stopped with all the will in the world. Diaphragm, belly and chest contract, and it's as if you suddenly had to hiccup really badly. The difference in pressure between your inside and your outside increases, and then the uppermost sphincter of the oesophagus springs opens. Like a tsunami the contents of your stomach force their way out. Shutting your mouth tight won't help; it only comes out your nose. All this violence is necessary because puke runs against the flow of traffic. Normally, the traffic in the alimentary canal is strictly a one-way street, from mouth to anus, from food to shit. You'd be dumbfounded if the turd you had just shat were to rise up out of the toilet, enter your lower intestinal tract and reappear again in your mouth bite by bite as steak with mushrooms. The reason the body has broken the rules is only because it hopes by so doing to prevent something worse. You puke in order to get rid of poisonous substances or to remove a blockage. To induce it deliberately all you have to do is ingest an emetic or stick a finger down your throat. The ancient Romans tickled their uvulae with feathers to keep their orgies going longer.

The most revolting thing about vomit is the sour smell. But that's not the only thing about vomit that's sour. If you forget to clean up your cat's puke it will burn a hole in your carpet. With a pH of 1, gastric juice is among the world's strongest acids. The second greatest mystery of the stomach is how it can tolerate all that acid; the greatest mystery is how glands can exist that produce an acid in which they themselves would dissolve. The secret lies in other miracle glands that produce a miracle mucus which keeps the burning acid away from the delicate stomach tissue. If these glands fail to deliver for one reason or another, you run the risk of developing an ulcer that may result in a perforation of the stomach if the ulcer festers. When that happens, the acid drains out of the stomach and ends up in organs that are absolutely defenceless against it because they have no miracle glands of their own. The acid is supposed to stay in your stomach, where it preserves your food until it's ready to enter the rest of the digestion machine. In this ruthless chemical warfare, bacteria decompose and pathogens meet a miserable death in the gastric juices.

And then what? Does anything else happen in the stomach? Vomit is silent on the subject. In order to find out, you would again have to arm yourself with gun and camera, or a gnome's cap. Or an endoscope, of course. But before the endoscope was invented, there was only one alternative outside of the gun, camera and cap: the break-in. If you can't get in by the front door and there's no visibility from the back, your only choice is to go straight through the wall. Operate. But all you get by operating is an arbitrary view. The only thing that really helps is a lucky break. Science needed continuous insight into human inner workings, and it was then that coincidence came to the rescue. With a marvellous accident.

On 6 June 1822, Alexis St. Martin became a man with insight. A musket ball had accidentally torn a hole in his stomach, and as a result everything that went into his mouth came back out through this hole. The doctor from Fort Mackinac, Michigan, kept him alive by means of 'nutritious injections' rectally administered. The patient recovered, albeit with an exit hole in his belly, which 'was a genuine anus, except for the absence of a sphincter'. For Dr William Beaumont, an ambitious army doctor without a war to practise in, this was the chance of a lifetime. Science finally had a man with a car bonnet. Beaumont tied pieces of food to a silk thread and inserted them into Alexis's stomach. In this way he could pull the thread out at regular intervals to see what had happened. But he went too far. At one point he shoved sixteen raw oysters into Alexis's belly; another time it was a thermometer. Shamelessly leering in, he found that Alexis's stomach turned red when he became angry. And he had every reason to be. Sick and tired of serving as a human guinea pig, Alexis put an end to the tests in 1832, the year before Beaumont became famous with his *Experiments and Observations on the Gastric Juice and the Physiology of Digestion*. But not everyone was happy with the discovery that you could reduce *vis vitalis* locally to an ordinary toilet bowl cleaner, hydrochloric acid.

Later on, hydrochloric acid was also found in the stomachs of sparrows, panda bears, nine-banded armadillos, frogs and toads. Even greater was the astonishment in 1973 in Australia when a new species of frog was discovered, *Rheobatrachus silus*, that vomited living tadpoles. It seemed that the animal hatched her larvae in her stomach. Biologists couldn't believe their eyes. It's hard to imagine an environment less hospitable for the raising of youngsters than a stomach, where food is prepared for digestion. Tadpoles are definitely not resistant to toilet bowl cleaner. The only possible explanation was that when the frog hatched her young in her stomach she temporarily halted the production of hydrochloric acid. But how?

Tests showed that the trick is not carried out by the mother but by the young themselves, back when they were eggs. The mother is totally oblivious; she probably regards her eggs as food, which is how they get in her stomach in the first place. There they secrete a substance, prostaglandin E_2, which suspends the production of hydrochloric acid. So the larvae fiddle with the body of their mother in order to transform a deadly witches' kettle into a cosy little cot.

People don't gestate in their stomachs. Our children have a room of their own, right in the parental body: the womb. That may seem convenient but it's also rather a waste, having an organ that is used for only a short part of your life and otherwise just sits there, useless and unoccupied. While we fill our stomachs three times a day, that other organ is filled an average of only three times during an entire human lifespan. The reason is clear. An empty stomach convulses, which causes an unpleasant feeling called hunger, and the only way to make that feeling go away is to fill the stomach. Wombs are never hungry; a womb doesn't growl. Satisfying the desires that go hand in hand with the filling of wombs requires a very different organ, a much smaller one, located a little further on.

The stomach also farms out the desires associated with filling it. That's what it has the tongue for. As icon of gastronomic gratification, the stomach is highly overrated. You can actually do without it entirely. People whose stomachs have been removed, or who have had a gastric bypass, live quite peaceful lives, and they're stylishly thin as well. Many species of animals have never had stomachs. We regard the stomach as the centre of digestion because of our need for hierarchy. If the heart rules over the circulatory system, then the stomach will rule over the intestines. But the heart is no more than a pump, and the stomach no more than a warehouse. Except for water, almost the only thing that is absorbed into your blood from the food and drink in your stomach is alcohol, so that wine or beer can soon be detected on your breath. Opinions are divided as to whether this is convenient or not.

The stomach is simply a broadening of the intestinal tract. At best you can see it as a single station on a long underground line that starts in the throat, dives through the oesophagus on the way to the stomach, climbs up a slight incline to the duodenum and, after a long and winding ride through the small intestine via the Caterpillar ride of the large intestine—straight up, along a horizontal bit, and then right and straight down—arrives at Anus Station, its final destination. One way only; no return tickets available.

The route's first stretch goes via the oesophagus and carries the food and drink through the chest, behind the heart and lungs, towards the abdomen. Upon its arrival, the stomach opens its gates, which otherwise are tightly shut to prevent the stomach acid from percolating upwards and incinerating the oesophagus or the throat. As a station, the stomach is really no more than a waiting room, although there are snacks available there such as pepsin to initiate the digestion of proteins (carbohydrates have already undergone a preliminary treatment in the mouth with the enzyme amylase). Within six hours the transport is ready to continue. No matter what delectable ingredients you put in your food, no matter how sophisticated your cooking technique, your body feeds itself with nothing but the vomit dripping out of your stomach and into your intestines. No photos of food porn here. The only time we see what's really keeping us afloat is when we have to throw up. Sometimes you recognise a tiny bit of chicken or caviar, and almost always chili con carne, even if you haven't eaten it in quite some time.

Before the vomit in your belly can really be digested, the gastric juices have to be neutralised. This is done with an equally strong antacid: bicarbonate. In the shop you can buy it as antacid tablets; in your body it's produced by the pancreas, which empties into the duodenum just below the stomach. Only when the last morsel of puke is sufficiently neutralised does the sphincter controlling the stomach's exit door release a new batch. This sphincter is known as the pylorus—'gatekeeper' in Greek—but in fact is no more than a back door that is operated at a distance by the real gatekeeper, the pancreas, with the help of the bicarbonate. The pancreas is also responsible for producing enzymes such as trypsin, lipase and amylase, which break down proteins, fats, and carbohydrates respectively. Fats are the most troublesome. They need a lot more than lipase to be broken down properly. First they're reduced to little droplets by the bile salts that the liver supplies to the intestine via the bile. But no matter how small, even the tiniest morsel of pork chop or Brussels sprout won't fit through the openings in the intestinal wall and has to sit out the entire journey unutilised, only to get out at Anus Station, its final destination. Strictly speaking, everything in the intestines is located outside the body, the way all the air in the hole of a spool isn't part of the spool but of the outside world. To break down the food particles so they can get through the intestinal wall, the intestine resorts to an old trick from the chemistry lab: it liquefies them. Solids are usually much more manageable once they're liquefied. As tiny, water-soluble molecules, the nutritious substances can hazard the crossing from the intestine to the nearby blood vessels. After arriving in the blood they often combine immediately to form complex, insoluble substances, but that no longer matters now that they're safely on the other side, in the actual body. Here, the ready-to-eat fragments travel via the blood and the lymph to the liver, which makes the building materials and fuel that we need in order to live. The actual building and burning of fuel then takes place in various cells all over the body. Each cell has its own power station, but it runs on raw materials that are supplied by the intestines.

In order to extract sufficient energy from your food you need long intestines. A human being has six to eight metres of them. The duodenum—which means 'in twelves' in Latin—makes up for twelve finger-widths of intestines, or twenty-five to thirty centimetres. After that come metres of small intestine before the large intestine spans the last one and a half to two metres.

With all their entrails, large animals soon run up against an intestinal shortage. The nutrients are absorbed through the intestinal surface. If an animal grows to twice its length, then its intestines attain a surface that's 2 × 2 = 4 times as big. That sounds good, because it means the animal can process four times more food. But the animal itself becomes 2 × 2 × 2 = 8 times as heavy. If an animal eight times bigger gets only four times as much food, it falls short by a factor of two. There's hardly any room to expand the surface by means of folds and bulges, since the surface has already been increased to the extreme. A human intestine, for example, has five hundred times as much surface area as the inside of a smooth tube, thanks to the intestinal villi; that's a whole football field. To keep from starving, humans have no choice but to let their intestines grow faster than their body. That's quite a squeeze. The intestines have to twist themselves in all kinds of convolutions to find enough space. In a human embryo the intestine starts out as a straight tube. Because it grows faster than the body in which it is housed, the intestine first makes a turn of 90 degrees anticlockwise, and then another turn of 180 degrees; it keeps on coiling in this fashion because the tissue to which it is attached at the back of the abdominal cavity doesn't grow fast enough. By the time you reach adulthood, your intestines are still hanging from your backbone by means of this mesentery; in the front they're held in by the stomach muscles. Except when they aren't. Then you get a paunch.

Intestines have always had difficulty keeping up with the growth of their body, not just within the span of a single human life but within the evolution of a small to a large animal species as well. Stuffed into such an awkward space, they start twisting of their own accord, much like a rubber band in an undersized bag. But it can't go on like that. At some point there's no more room for twisting, and the animal has to adapt its lifestyle. This explains why cows can't fly and birds can. It isn't so much a question of wings as it is of intestines. Cows eat grass. There's not much nutrition in a single blade of grass, so the cow has to eat a lot of it, which its massive amount of intestines also demands. They almost bulge out of its body. A cow is built around its digestive system. This gives it that ponderous, bony, old-fashioned hay-making machine look: in short, that poignant quality. But flying is out of the question. Birds fly, though, even if all they get to eat is grass. A wigeon duck, for example, eats nothing but grass, yet still it flies. Like a cow, it has unusually long intestines for this type of animal, but when they reached 130 centimetres the wigeon decided that was good enough for a bird. If its intestines had grown any longer it would have had problems. In aviation, there's a heavy price to pay for excessive baggage in terms of fuel. To get the extra intestines airborne, a wigeon would have to eat more grass, for which it would need more intestines, etc. Somewhere there's a golden mean between intestinal weight and flying capacity. That compromise is called 'wigeon'.

At the end of the small intestine the food sludge reaches a fork in the road: go straight to the appendix or take a left and enter the large intestine. The only thing the human appendix can still do well is become infected and cause trouble. Like a murky alley it comes to a dead, worm-shaped end. So all the sludge moves on to the large intestine. There's no turning back now; that option was ruled out for good at the T-junction by the ileocecal valve. At this point the food is as good as digested, and all that's left of your meal is a watery sort of pea soup. Now it's the job of the large intestine to make lovely turds out of it. This is mainly a question of thickening, which means the water has to be drained off. So it gets carried away through the intestinal wall along with all sorts of salts that are of use to the body. Every day, about a bucketful of water is removed. If you didn't have a large intestine, you'd have to drink that bucketful every day to make up for it. It's not always possible to find that much water, which accounts for the development of the large intestine during evolution: it makes it possible to live in dry climates. This requires life's cooperation, however. The pumping station of the large intestine does heavy work due to the salt content. Once the salts are pumped to the other side the intestines absorb water automatically. But all that salt pumping takes a lot of energy, and therefore a lot of fuel.

Otherwise there's not much more for us to do in the large intestine. We find ourselves in someone else's territory. The large intestine is the domain of the microbes. For them, it's a bustling metropolis full of nightlife, gangs and junk food. They eat what our enzymes in the small intestine aren't keen on. While we need oxygen by definition in order for our metabolism to function, bacteria also thrive in the oxygen-free backstreets of the large intestine owing to the fact that they can ferment their food, a chemical process requiring no oxygen and better known as rotting. It can be relatively stinky. You only notice this when you fart. You might find farting embarrassing, but sensible people ought to be proud of this sign of life from all their little co-workers. Without them we would have been extinct ages ago.

With a hundred trillion bacteria, the large intestine accommodates ten times more cells than we have in our entire body. The only reason we aren't weighed down by them is that bacteria cells are so much smaller than the cells in any random organ. All of them together weigh a kilo and a half, not much more than our brains. There are about a thousand different species of intestinal bacteria. Fortunately most of them have developed a friendly relationship with us over the course of evolution—until we start taking antibiotics. During a typical treatment half the species give up the ghost in an enormous massacre, which prevents the intestinal flora from doing their work. Not only do they process our leftovers but they also provide vitamins, protect us from pathogens and stimulate the immune system. What you ought to be doing is cherishing them. Instead of antibiotics, many doctors prescribe intestinal bacteria to fight allergies, obesity and vascular disease. If your own microbes are falling short, it's time to call in the auxiliary troops. For a push in the right direction, many people take probiotics such as Yakult or Actimel. Sceptical doctors doubt whether the friendly bacteria in these probiotics can really pep up lethargic intestinal flora. They see more benefit in the transplanting of healthy faeces, which can be necessary when the virulent hospital bacterium *Clostridium difficile* seizes the opportunity to stage a takeover in a large intestine after a course of antibiotics. You can try hosing your intestine down, but *C. difficile* will just rear its head again—unless you inundate it with healthy faeces from someone else's intestine, which can be done by inserting a tube in the nose and running it down into the small intestine. Fortunately there's little reason to take such drastic measures with every rumble in your tummy. Such gurgling is not meant to alarm you but to reassure you; it's a sign that your intestine is being properly looked after.

Of the thousand species of intestinal bacteria, every individual has about 150 in an entirely personal blend, each person with their own little garden. Tell me which bacteria you have and I'll tell you who you are; an intestinal profile is as specific as a fingerprint. Yet as an embryo you didn't even have one species of bacteria. An embryo enters the birth canal in as sterile a state as urine in the urethra. It isn't until birth that a baby acquires its first portion of intestinal flora. Every vagina is a microbial paradise, and the composition of the flora is a reflection not of the newborn but of the mother. Unless the baby came into the world by way of Caesarean section, in which case it's a reflection of the nurse who first handled it. But that would be skin flora, which differ from every kind of intestinal flora anyway.

In order to stay on good terms with your intestinal flora, you provide them with a steady diet of fresh intestinal mush. That's not easy. How do you propel something along a conveyor belt many metres in length without wheels? Intestines have muscles, not wheels, and they can only do one trick: contract. With the circular muscles the intestine makes itself thinner, with the longitudinal muscles it makes itself shorter. When the circular muscles squeeze together the contents squirt from one side of the contraction to the other. There should be valves, as there are in the heart, which neatly press the blood to the other side. The small and the large intestines have only sphincters at the extreme ends. In the sections in between, movement is achieved by means of peristalsis. In peristalsis, the intestine doesn't squeeze at once over its entire length but muscle by muscle, so the constriction runs through the intestine like a wave, rather like when you squeeze the last bit of toothpaste from the tube. While one circular muscle squeezes the ball of food forward from behind, the circular muscle in front of it relaxes to make it easier for the food to pass through. The longitudinal muscles pitch in at just the right moment to give the circular muscles a hand. In this way the intestinal contents move along at a good pace. After a stopover of a few hours in the stomach, the food takes about an hour and a half to get through the small intestine. The small intestine doesn't care if the mush is thick or thin, but fatty food slows down the action because it pinches off the flow from the stomach. But even before digestion is complete, the muscles sweep the small intestine clean every other hour and a half to keep bacteria from settling there. Compared with the large intestine, the small intestine is practically sterile. It takes longer to get the contents through the large intestine than the small if only because its diameter is about twice as large. The fragments get tougher and tougher, and in order to keep them moving the large intestine has more powerful circular muscles than the small intestine and its inner wall is better lubricated with mucous. But it isn't only up to the muscles to decide when to let the shit see the light of day. The large intestine also serves as a storage depot, waiting until the coast is clear and it's safe to dump. The muscles can't tell when that might be.

The tempo of one trajectory has to be coordinated with that of the other. The contractions cannot be allowed to collide. And not only do the contents of the intestine have to be pushed forward, but they also have to be kneaded and mixed. The circular muscles have to open and close on time, fluids have to be released on schedule and according to the right dosage, and bacteria have to be put to work. Compare it to brewing beer or baking bread; one mistake and your dough is ruined. So digestion makes heavy logistical demands, which call for scrupulous guidance and control. But from whom? You yourself wouldn't know how to begin. Or do you know exactly when the gastric juices are sufficiently watered down, when to add more bile, when the duodenum has to be urged to speed up, when a turd is thick enough?

Are you in the driver's seat, or are you just nothing but your own car? We drive our bodies the way we drive a car: we know where the steering wheel is and the gas pedal, but most of us have no idea how the engine works. Open the bonnet and you glaze over, look inside yourself and you're nauseated by your own bowels. Recently something inside me broke down and the doctor invited me to look at the monitor with him so I could see inside. I saw my own bladder. Nothing about it looked familiar. Fortunately there wasn't much damage. I gave my bladder a good talking to and I hope I never see it again. I'm perfectly happy to leave it to the competent authorities who know something about the autonomic nervous system.

This system has traditionally been divided into two parts: the parasympathetic nervous system and the sympathetic nervous system. Eating and digestion are the work of the parasympathetic division. This is the nervous system of repose. It slows down the heartbeat, the breathing and lots of other stuff. The only things that are stimulated are the appetite, the pumping of the intestines, and the sex drive. It's the fun nervous system. Unfortunately the sympathetic nervous system gets in its way. It wants to turn all of us into scrawny managers with ulcers. Sensible managers know how to switch off this nervous system, however: with a business lunch or dinner. Like a coin in a vending machine, these kinds of meals switch on the fun nervous system so you can calmly get down to business. At least that's how I learned about it as a student.

In the late twentieth century this picture was drastically overhauled. The running of the stomach and intestines was declared independent. In and around the intestinal tract there are hundreds of millions of nerve cells that do everything on their own initiative. Within the autonomic nervous system they form a kingdom all their own, in addition to the sympathetic and parasympathetic nervous systems. This is where the sense of rhythm is to be found that keeps the intestinal muscles dancing; this is where extra bile is delivered; this is where the decision is made that you're hungry. This enteric nervous system is what neurologist Michael Gershon called 'the second brain' in his book of the same name. Thanks to this second set of brains, the supreme brain high in our heads doesn't have to interfere with our intestines way down south, thus freeing up its cells for more elevated things. But what do we mean by *elevated*, Gershon wonders:

> The enteric nervous system may never compose syllogisms, write poetry, or engage in Socratic dialogue, but it is a brain nevertheless. It runs its organ, the gut…When the enteric nervous system fails and the gut acts badly, syllogisms, poetry, and Socratic dialogue all seem to fade into nothingness.

Even though the second brain is connected to the first brain by means of a main cable, the vagus nerve, so the first brain can oversee what's going on, if you were to sever this nerve connection the intestines would still go on working. It's more likely that any problems would occur in the opposite direction. Far more signals are sent from the belly to the head than the other way around, as if the intestines were telling you what you ought to do. It's just a matter of how you look at it. Seen from the vantage point of the first brain the intestines serve us as a power station, a typical public utility. But if the second brain were to have any capacity for self-consciousness, it would know that life is not based on philosophy but on physiology. We *feel* ourselves with our first brain, but we *are* our intestines with their second brain. The fact that our brain is only a walnut-shaped appendage of our intestines takes some getting used to, but when all is said and done it doesn't have a lot to contribute. The second brain has a powerful weapon to get our first brain to do its bidding: hunger. If it came down to a direct clash between intestines and intelligence, then the intestines would win, as we see from the report by stomach-intestine specialist Akkermans on attempts to tame a stubborn bit of bowel:

> When a large section of the small intestine is removed during an operation, the rest of the intestine may be too short to absorb sufficient nutrients. To deal with this problem, surgeons cut away a piece of small intestine, turned it around, and then re-attached it in its original place, only backwards. The direction of the peristalsis in this piece of intestine would then be the reverse of the peristalsis in the rest of the intestine. As a result, the speed at which the food components were transported through the intestinal lumen would be inhibited, leaving more time for the absorption of nutrients. This trick worked for about a month. The neurons in the enteric nervous system grew out through the connecting seam and the peristalsis in the implanted piece of intestine reversed itself.

Akkermans's conclusion: 'Apparently it is difficult to fool the enteric nervous system.' But it may be possible to drive it crazy, according to Michael Gershon. 'Since the enteric nervous system can work independently, we should consider the possibility that the brain in the belly also has its own psychoneuroses.' This raises hopes for the treatment of nervous disorders such as irritable bowel syndrome. Maybe the brains on the psychiatrist's couch should be those in the belly and not those in the head.

Just as the underground railways in all the world's cities look alike, so the intestines of all animals are made the same way. The food enters in the front, releases its energy along the way, and at the last station everyone is kindly asked to leave. The senses keep a close watch on everyone when they enter, but there's hardly any checking at all at the exit. Yet one city isn't quite the same as the next, and there are also significant differences among the various animals. One is bigger than the other, or more active, faster, more modern. Large animals have long intestines, but those of a lion are shorter than those of a gnu, who's just as big. The intestines of a lion have less to do because it's a carnivore. Meat is easier to digest. It consists of animal cells, which have thin walls, so you can easily get to the contents. Plant cells are wrapped in thick capsules of cellulose, which is difficult to digest. First you have to crack them open like nuts. Babies can't do that at all. That's why every human being, vegetarian or not, must begin life as a carnivore. Not with a lamb chop or a hot dog but with mother's milk, which is a pure animal product. This white, liquid meat is full of amino acids that a baby would never be able to extract from a vegetarian diet on its own. By nursing, every mammal has at least one carnivorous period. Birds don't nurse, although doves feed their young a milk-like substance from their crops. But in many cases the species that eat sunflower seeds like proper vegetarians and dutifully cling to the peanut baskets hanging from trees are the same ones that grew up on nutritious caterpillar mush.

While a carnivore quickly jams its tender prey through its short intestines—a cat has to return to its dish within thirteen hours—a herbivore needs more time to get the job done. Big herbivores take from forty to sixty hours. Is that why vegetarians stay so thin? Or do they only appear to be thin? Thin, not to say gaunt, is the general impression that the butcher's regular customers have of those who abstain from meat. But Bernard Shaw, Leonardo da Vinci and Greta Garbo aren't the only well known vegetarians. The hippopotamus, the elephant and the bison are just as horrified by meat. Yet they're all certified heavyweights. In order to extract enough energy from its paltry fodder, each hippopotamus has fifty-five metres of intestine continuously running full tilt. A lot of the energy produced is needed right away for the kneading and pumping of the next serving of plants. And it takes a great deal more energy just to lug all those guts around, which means having to eat even more. All the livelong day. In order to chew all those plants, the head of a herbivore is almost entirely taken up by jaws, molars and chewing muscles; there's little room left for brains. Predators, on the other hand, need a lot of intelligence in order to overpower their prey by means of cunning and deception, strategy and deliberation. Herbivores use their wits too, to stay out of the clutches of the carnivores, but they're less successful; otherwise the carnivores would soon die of hunger. Eating meat makes you smart. It's fun being smart. That's why we have warmer feelings for cats than for rabbits. Carnivores are more playful, more intelligent, and more human-like than herbivores. Nicer. The only thing vegetarians have going for them is being right. And unfortunately you can't eat that.

For grass-eating animals, a long intestine isn't enough. They need extra bacteria to ferment the tough cell walls. To keep these auxiliary troops from being defecated before they're able to do their work properly, the intestinal underground has built additional stations for them. With a cow these are the four stomachs. Here the bacteria break into the cell walls before the grass can get to the small intestine, which soaks up the nutrients. To help the bacteria along, cows vomit up the contents of their stomach every now and then so they can chew it some more. In this way they put their leisure time to good use (when they aren't grazing) by maintaining their energy supply with their jaws. For all the work the bacteria do, they're fobbed off with 6 per cent of the total extracted energy. The efficiency of the system is reflected in the huge milk yield as well as in the squishy cow pats, which contain few nutrients. A horse does it differently. Instead of four extra stations it only has one, a three-part room en suite that comes after the small intestine rather than before it. This is where the large intestine is made ready for the bacteria. Hindgut fermentation like this is not as efficient as the foregut fermentation of a cow. That's too bad for the horse, but it does result in lots of very fine horse droppings.

The small intestine isn't the best place for fermentation because bacteria aren't very good at dealing with enzymes, and the small intestine is full of them. So which is better: letting your plants ferment before or after the small intestine? That depends on the available food. In a juicy Dutch meadow it's best for the plants to ferment first, in which case you become a ruminant, like the cow. But if grass is scarce or straggly you'd be better off as a horse, because then you don't have to chew your cud and you can spend the whole day grazing. But what if you're a rabbit? Even though a rabbit's food is more nutritious than mere grass, it still needs an efficient digestive system because it's so small. Small animals need more energy, relatively speaking, than large ones. Once it was thought that rabbits chewed their cud. The endearing way they have of wrinkling up their nose was taken for chewing. That almost cost the rabbit its head. According to the laws of ritual purity in the Authorised Version of the Bible, ruminants were fair game. Only the fact that the rabbit does not have a cloven hoof saved it from the Old Testament cooking pots. Only later was it discovered that in the Bible the word 'rabbit' was a translation error. While the Authorised Version speaks of rabbits, the Hebrew has 'hyrax', or dassie. Hyraxes may look a little like rabbits but they live like marmots, are related to the elephant, and move their jaws like ruminants, although they don't chew their cud. Rabbits aren't big on cud-chewing either. Their food is fermented like that of the horse, following the small intestine—not in the large intestine, however, but in the appendix. The appendix in a rabbit is a sizeable, vital organ, just as it is in many of the apes. We humans almost lost our appendix over the course of our evolution because we went from being herbivores to being omnivores and could easily live without it. For a rabbit, on the other hand, the appendix is essential. In addition to nutrients, a special kind of faeces is made there, softer than ordinary rabbit droppings, darker and more strongly scented. You never see these pulpy droppings out in the fields, and for a simple reason: rabbits eat them as soon as they've shat them, nice and juicy and covered in a membrane, straight from the anus. In this way the rabbit profits from the nutrients, vitamins and bacteria that would otherwise be lost. It really does resemble chewing the cud in that sense. Just as in cud-chewing, the extra energy is released mainly when the animal is at rest, giving it a 10 to 15 per cent energy boost.

As an omnivore, a human being can manage with a simple intestine. Human food is rich and digestible enough so there's no need for cud-chewing, re-pooing, or foregut and hindgut fermentation. And in terms of length, a human intestine—which is longer than that of a carnivore but shorter than an herbivore's intestine—is just what you'd expect of an omnivore. Unless you were Dr John Harvey Kellogg of cornflakes fame. He believed that with all that cooked food, human intestines weren't as active as they had been in prehistoric times and it wouldn't hurt to cut off a bit. Otherwise they just kind of hung there. Food got stuck in them and began to rot. The rottenness entered your bloodstream, and before you knew it you were pushing up the daisies. Fortunately, you could have your intestine shortened in the Mecca of American food fanatics and religious lunatics, Battle Creek, where Kellogg ran a sanatorium. Following in the footsteps of surgeon Arbuthnot Lane, who cut away substantial sections of the large intestine in order to give it a boost, Kellogg himself took scalpel in hand:

> Of the 22,000 operations I have personally performed, I have never found a single normal colon.

Only later did it occur to Kellogg that not only can you adapt your intestines to the food you eat but you can also adapt your food to your intestines. That saves a lot of blood. His brother Will began running a factory for making special food for the large intestine, and cornflakes were born. In no time at all, cornflakes had driven ham and eggs from many an Anglo-Saxon breakfast table. To this day you still have to be careful about where you spend the night in America or England. Before you know it there's a bowl of milk-soaked kibble staring up at you. The same thing is threatening to overtake the Dutch. Never have I understood how someone could trade a splendid breakfast food like coloured sugar sprinkles, or syrup that you can write your own name with, for chicken feed. Until I read Dr Kellogg's loathsome little books. Finally I get it. Cornflakes aren't for eating. They're for shitting.

5

Do-It-Yourself

Humans are makers. They take pride in it. *Homo faber.* See them beaming beside their successful homemade cakes, the chicken coops they built themselves, their own begotten offspring. The writer eagerly plants his signature on the latest way to line up a hundred thousand words in a row. *His* way. Another addition to the structure of reality. A tower, a legislative bill, a football goal, a quarrel: humans make them. And all those creations make us human.

But the eagerness with which humans show off their handiwork is more than equalled by the secrecy with which they treat the speciality of their backside. Yet it's a wonderful product, the most homemade of them all, a top-notch solo performance. You'd be tempted to take out a patent on it if you didn't already know how little demand for it there was. The first person to see commercial value in it nevertheless was the Italian artist Piero Manzoni. In May 1961 he sold his faeces at a local gallery in

30-gram tins, neatly labelled in four languages as *Merda d'Artista* (artist's faeces). Each of the ninety tins was numbered and signed. One of them was auctioned at Sotheby's in 2007 for 124,000 euros—three thousand times more than the fresh merchandise cost in 1961. Even your most individual faecal statement doesn't bring in that much. Do artists really defecate so much better than you do? Of course not. But you have to be an artist to elevate a turd to the level of art.

As far as that's concerned, anyone can call himself an artist. A turd is the most democratic of all works of art, but that doesn't make it the simplest. Compared with the miracle of creating a full-blooded piece of excrement from a hamburger with fries, the transubstantiation of bread and wine into flesh and blood on the altars of the faithful is a magic trick for children's parties. Go ahead and try it. Bring home a hamburger with fries and you'll end up with an honest-to-goodness homemade human turd. Throw in a kitchen full of pots and pans, the most expensive chopping and kneading machines, *Escoffier* and *Julia Child*, but without the mechanics of our entrails you won't get very far. Even the top of the top chefs has to acknowledge the superiority of his own intestines.

Can shit be made without intestinal intervention? Nothing could be easier, or so you would think. Progressive schools and mischievous uncles encourage children to make turds out of moistened gingerbread. The result is just as feeble as the joke products from the party store. Nappy manufacturers who want to test their products under odourless conditions resort to substances like mashed potatoes, peanut butter or pumpkin pulp, but that only makes the muck worse. The American firm Kimberly-Clark switched to fake faeces made of cellulose powder, wheatgerm, resin, colouring agents and water. In Japan, Toto tests its toilets with artificial turds made of miso paste, a mixture of fermented soy beans, rice, and salt. Here the work of our intestinal bacteria is replaced by yeast. But it's still pretty rudimentary.

It took until the turn of the century before an artist produced artificial faeces with the help of intestines—albeit artificial intestines. In the year 2000 Wim Delvoye (1965) launched his *Cloaca* in an effort to imitate digestion. The machine consisted of six large retorts, connected by tubes and computer cables. It reproduced the processes that take place in your body. Three times a day, food prepared by a top chef was fed to the contraption via a funnel. By way of thanks, it shat a well-formed turd every day that in many respects could hold its own against one of yours or mine. For roughly a thousand dollars art lovers could take a sample home with them, not in a tin but in transparent plastic, so you could be sure you were getting your money's worth. It still seems expensive, but the Belgian artist did achieve something that no one before him had succeeded in doing. A turd from *Cloaca* wasn't waste material—collateral damage. It was the actual product. That's what it's all about. Unlike many other scatological artists, Delvoye didn't just make art from shit; he made shit art from artificial shit. The success of the *Cloaca Original* led to a series of new models in which the tempo of a single digestive cycle was stepped up from forty hours to six, making it possible to coordinate the completion of the cycle with the gallery's opening hours. Fortunately, many of the models have retained human characteristics. From time to time a Cloaca will release farts through its exhaust pipe, and occasionally it gets the runs. Probably something it ate. The botched turds are not sent to the rubbish bin but are properly transported to the toilet by a gloved assistant and flushed away.

How do you make a good turd? With care, experience, love, pleasure, and of course with intuition. But what you mainly need is luck. If your luck is bad, nothing will turn out right, no matter how hard you try. It's just as if you had no influence over the process at all. Cookbooks can really let you down. All they're interested in is the pleasure you enjoy immediately after every bite, so that the ultimate satisfaction loses a lot of its lustre the following day. Rarely, if ever, do restaurant reviews report the long-term effects of a particular meal. A restaurant can easily get three stars from Michelin for a menu with such an unhappy ending that yesterday's gourmet finds himself with an anal hangover today.

Many a dinner party has ended in a disappointing little turd. Cooks don't take our intestines into account, just as intestines weren't made with cooks in mind. Cookbooks are collectively responsible for the bad smell of human faeces. Pans and intestines compete with each other. Pans take over the work of the intestines to a large extent. Cooking is predigestion. Instead of tough meat and indomitable turnips, the intestines are served a predigested mush consisting of partially broken down muscle protein and plant cells that have been boiled to death. When cells come in contact with boiling water and hot fat, their walls burst open and the tightly entwined protein chains are loosened; the resistance of the tough tendons in braised beef is broken by acidic ingredients and simmering. What ends up on our plates is as predigested as the dollops that nestlings are fed, vomited into their beaks from the crop of their mother.

If you're looking for the ideal excrement, the artist or the cook are not the ones to consult. Go instead to the butcher. Butchers make sausages. Once you get them past your molars, sausages have little to offer; the turds they produce are just as nondescript as those that come from other meats. But when they're still in the shop, the racks full of sausages resemble a festive prelude to a Miss Turd Pageant. They hang there gleaming with pride, stuffed to capacity, and hoping for a trophy for the best mettwurst, a gold medal for salami, an honourable mention for genuine Thüringer Bratwurst. Amidst beer and obscene comments the prizes are awarded. The gratifying results later appear in the village newspaper, and the neighbouring pigs are less certain than ever about their future. Yet making a prize-winning sausage is not the most challenging job for the butcher. You learn to make sausage at the butchers' vocational school; all the necessary ingredients are for sale. In their own toilets, however, butchers are as much amateurs as everyone else. No diplomas hanging on those walls. You either can or you can't. But when it comes to the theory the butcher is definitely at an advantage. The physical resemblance between sausage and turd is not accidental.

How do you make a sausage? The same way you make a turd: with waste. For a good sausage you need good ingredients: remains of muscles, bits of lung, scraps of liver, blood, brain fat, everything the butcher deems too distasteful to be sold and too expensive to be used as dog or cat food. As the old saying goes (indelicate as it is accurate), buy a sausage or take a widow for a bride, you never know what's been stuffed inside. Good waste is still an oxymoron. To make it easier to process, and to hide it from overly sensitive eyes, the remains of the cadaver are ground up by the butcher with the help of a mincer or a cutter. This is the first step in artificial predigestion, and it saves our teeth a lot of work.

After spices are added, the discarded meat is mixed and kneaded into a mass that already bears a suspiciously striking resemblance to the contents of the turd it will eventually become. What's in it? A turd is made of shit. Shit has to meet the same kinds of requirements that sausage filling does. You have to be able to squeeze it out of the same kind of tube that sausage filling is packed into. The most striking resemblance between the two is the colour. In both cases there's something objectionable about it. To make things easier, these nebulous, rather gloomy tints are summed up in the word 'brown'. In many respects brown is the lowest of colours. In practice, brown is usually what you get when you rinse paint brushes used for several different colours in the same pot of turps. Even the cheeriest of colours are no match for brown. It's a colour graveyard. What this says about those rare periods in history when the colour was in fashion—the 1970s, the second half of the nineteenth century—is something I'll leave to the cultural pessimists.

Brown is offal. 'Offal' refers to animal waste, the chopped-up entrails and internal organs that was the food of the poor in earlier times (hence the expression 'eating humble pie'). Today the butcher uses offal in sausages, which accounts for the poo colour. The brown of both shit and sausage began as red. Shit owes its colour to red blood cells, whose job it is to circulate oxygen with the help of red haemoglobin. There's hardly any blood in sausage (with the exception of blood sausage), so there's little red blood pigment. But when the meat in the sausage was still muscle tissue it was also responsible for transporting oxygen. Instead of haemoglobin, a muscle uses myoglobin, which becomes just as red in the presence of oxygen, and for the same reason: the active part of the molecule that binds oxygen, the haem group, makes it turn red. This is where the iron can be found that you have to eat to keep from becoming anaemic.

Blood cells don't live forever. After more than a hundred days of tearing through the body, absorbing oxygen and releasing it, absorbing and releasing, absorbing and releasing, they get old and tired of life and are finally broken down by the spleen and the liver. Usable parts, like iron, are processed into new red blood cells, and the remaining waste, along with the bile, is sent on to the small intestine. The red colour is gone by then. In the small intestine the waste turns yellow from bilirubin. Bacteria convert it into urobilinogen, which oxidises into stercobilin and gets darker and darker as the concentration increases, until it leaves the body. But it could have been even browner. A lot of the bilirubinogen is transported to the blood early on through the intestines and put back into circulation. So some of it travels through your liver, bile and kidneys and ends up in your urine. The yellow in your urine comes from the same substance as the brown of your faeces, although (to make things more complicated) it has a different name: urobilin.

If all goes well, a sausage is completely eaten before the myoglobin is broken down into its constituent parts. Yet it doesn't take long for a slice of sausage to stop looking fresh. While a healthy, living muscle may be a cheerful red or pink thanks to the presence of oxygen, once it leaves the cow or pig and ends up as dead meat in the sausage only the exterior molecules continue playing the old trick of making red oxymyoglobin from purple myoglobin, and then only for a short time. Inside the sausage, the molecules use the small amount of oxygen that has worked its way in to make their iron atom rust, with the expected result: rust brown.

Brown is the colour of decay and mortality. Bruise a tree leaf and it turns brown, just like withered flowers, rotten apples and bad teeth. Substances that are no longer responsible for the continued existence of leaf, flower, apple or tooth, are broken down into complex compounds that absorb light from every possible point on the spectrum, and in so doing they reflect only brown light. All that death ultimately ends up in the earth, which consequently is also brown. And although all of life emerges from the soil—the green leaves from the brown branches, the green alpine meadow from the brown mountain—the first thing we associate with brown is death, the first thing we associate with the earth is the grave.

You can't go any lower than brown, which hasn't done much to raise the prestige of shit. How different the world would have looked if we all shat sky blue! Every now and then (certainly not every day) it would be nice to produce excrement of a more festive shade. To get this result you'd have to mix the necessary pigment with your food, pigments that don't lose their colour along the way. For a celebratory red turd you'd eat red beets with crimson berries, washed down with red wine. Consume the meal one or two days before the party, to give the pigment the required head start. You get yellow turds from rhubarb. You can also use senna leaves, but don't forget that senna is a laxative.

Black shit is easy to achieve, with the help of activated charcoal. I've produced white shit only once, after a lower GI series using liquid barium sulphate for contrast. Yet when I was a child, white shit seemed perfectly normal—for other people. If white people like me pooed brown, then brown people must poo white, or so I thought. Now I know better. Birds often shit white. This is due to white crystals full of uric acid, by which a bird is able to get rid of its nitrogen with minimal loss of water. Penguins alternate their white shit with a cheery orange variety, which comes from a heavy diet of lobsters. Flamingos themselves turn orange or pink from eating lobsters; penguins apparently don't want to attract too much attention in the middle of all that white snow and ice, and they excrete the lobster pigment in plenty of time.

Even bears reveal what they've been eating in the colour of their excrement: blue after blueberries, black after meat and brown after acorns. This results in a whole palette over the course of a year. Artist Gary Blum made use of this, but with geese.

> I collect the droppings, dry them in a toaster, and then crush them into a powder. When making the pictures you can't be too timid. I do get my hands right in there, but after the poop has been baked, it really doesn't smell. Each picture has four different colours of poop, which vary depending on what the geese eat. Dark brown is usually available in October-December, and is the result of the geese eating the tender roots of plants after the farmers have turned over the fields. The lighter colours are usually the results of eating grains.

Since 1994, when Goose Poop Art came on the market, Gary Blum has sold thousands of pictures.

In daily life, there are subtle variations in the browns found in chocolate and coffee. In the case of shit, too, the colour has a lot to do with the amount of milk that's been added. If you want to know whether there's sugar in your coffee you have to ask, but coffee with a lot of milk, or a little, is easy to spot. Milky shit is more yellow than pure shit. That's why baby poo is so light. Marie-Antoinette was so enthusiastic about the poo produced by her son, the dauphin (the crown prince), that she launched a whole yellow-green fashion under the name *caca de dauphin*. Later, many an English country estate was painted this poo colour by order of the National Trust in order to give it an authentic French look. As people grow older, the colour of their excrement changes along with the alterations in their eating patterns. Vegetarian poo usually stays light brown, but carnivores intensify the brown of their own discarded haemoglobin with that of the remaining myoglobin from sausage and steak. Lovers of blood sausage even manage to produce black turds with the help of the alien blood they consume.

In The Garden of Earthly Delights *by Hieronymus Bosch (1450–1516) the fruits of the intestines are plucked like roses.*

One big difference between sausage and shit is the smell. A pinch of spices may be enough to mask the sickly stench of the meat scraps in the sausage, but it won't work in your faeces, even though there are spices that can travel through the intestine without losing their fragrance. Thanks to the tenacity of allyl sulphides and vinyl sulphides you can always smell the onions or garlic you've been eating when you go to the loo. The odours of some foods fade, while others intensify. No matter how delicious the food may have smelled, once it reappears to greet the light of day it stinks to high heaven. The reason for this is that your intestinal bacteria have salvaged the amino acid tryptophan from the proteins in your food, which further decomposes into the notorious skatole. Skatole is the substance that makes shit smell like shit, and even makes you recoil from a turd that has long since been flushed away. Legend has it that the researchers who untangled the chemical structure of skatole in 1877 had to work far away from their laboratory colleagues, and that their white lab coats were later incinerated. An impact like that almost has to have been deliberate. Skunks actively repel attackers with their stench, and in much the same way our faeces keep other people at a distance in order to discourage the spread of disease.

As in the case of a volcanic eruption, it's the accompanying gases that make the biggest impression, no matter how much the shit itself may stink. That's because it's easier for gases to find their way to your nose from far away. The most feared gas is hydrogen sulphide, better known from stink bombs, which is distilled from sulphurous amino acids by the intestinal bacteria. The worst stench can be avoided by eliminating beans and cabbage from your diet and cutting down on meat, since meat contains a lot of tryptophan. Lowering your consumption of alcohol and pepper also helps. Alcohol and pepper throw your intestinal flora into overdrive, so they ferment even more rapidly. As far as odour is concerned you're better off fasting, in which case you end up shitting mainly intestinal pigments and other bodily residues. They don't smell bad. But in doing so you strike at the very heart of a turd. '*Was wäre Dreck,*' Luther wondered, '*wenn er nicht stinkt?*'—What's filth if it doesn't stink?

What kind of sausage is it? Does the turd meet every expectation? You don't know that until the sausage has been sliced, or the turd shat. Is it too hard, too soft or just right?

If sausage filling is too thick you won't be able to stuff it into its casing. The preferred material for such a casing is a real intestine, which breathes and stretches along with its contents. This intestine may come from the same animal that produced the meat scraps. Thus in the past—and to add insult to injury—many a pig was stuffed up its own backside. These days, most pigs are permanently separated from their entrails right after being slaughtered. While the actual animal is being processed into meat, its intestines are washed and sent on to the sorting plant. In China there are enormous factories that process ten million kilometres of pig intestines a year, good for about five million kilometres of sausage. Sheep intestines are 2.5 centimetres in diameter, only half as thin as pig intestines, but they make much better frankfurters. Back in the country of origin, the intestines are filled with scraps from a subsequent generation of animals. Naturally they don't have to come from the same family, and often they belong to an entirely different species, an intimate entanglement that suggests a kind of bestiality. And then you have intestines that are filled with intestines, but for such a perversity—the *andouillette*—you have to go to France.

Ultimately, the pig in sheep's clothing, or the sheep in pork casing, reaches the casing of our own human intestines. Quickly stripped of their sausage skin and surrounded by nothing but human bowels, the meat scraps inch their way through the small intestine, the large intestine and the rectum until they reach the exit. Now the big difference between excrement and a real sausage comes to light. A turd makes its appearance stark naked. If a real sausage were to do that it would immediately fall apart. Without an outer casing, soft sausage filling loses its cohesion, unless you've fried it to form a hard exterior crust. Then you end up with a croquette or a meatball. But that wouldn't work in our bodies. If you were to fry the turds in your body you'd fry your intestines along with them, and that would destroy them. The temperature of a healthy body is never more than thirty-seven degrees Celsius, so that's the temperature at which a turd is served up. Consequently, a turd reflects the temperature of the body's inner sanctum. That's actually the normal method for taking an elephant's temperature; inserting a thermometer into a fresh elephant dropping is easier than inserting it into one of the animal's bodily orifices.

Without crust or casing, a turd depends entirely on its own cohesion after leaving the body. But while it's easy for butchers to add less water to their sausage filling, there's almost nothing they can do to thicken the content of their own faeces. The tap is located in their insides, where they can't get at it. Actually, the reabsorption of water in the intestines is mainly regulated by the tempo of the conveyor belt. If the belt speeds up, there's less time for the water to be absorbed and the end product tends towards diarrhoea. The term 'the runs' is particularly apt in this regard. Excessive speed is usually just what is called for, since contrary to what is generally assumed, diarrhoea itself is not the problem but the solution. If the intestines think the wrong thing has been eaten, they'll want to get rid of it as fast as they can, at the cost of consistency. A great deal of water has to be removed to produce a good firm turd. But a minimum of fluid does not in itself mean maximum cohesion. Filling that is too dry just crumbles like sand. The particles have to stick together. That's why butchers add a binding agent to their sausages or stir an egg into the mince. In their own intestines they don't have to add anything at all. The intestines make their own binding agents from the products resulting from the breakdown of proteins in our food. Lubricants are also produced, to keep everything moving smoothly and to prevent disintegration. But thickening, binding and lubricating alone won't get you very far. You still run the risk of ending up with a long tube of brownie dough that gums up the whole works.

Wouldn't we have been better off with a casing or crust around our turds? The shit produced by young songbirds comes out in a little sack, which is tidily carried off by the parents. Elsewhere in the animal kingdom, among the insects and the crustaceans, the principle of casing and crust has been a great success. Inside its armour, a beetle or shrimp is as safe as a snail in its shell, a sausage in its casing, a croquette in its crust.

But that wouldn't work with an animal as big as a human. If our skin was all that held us together, we'd implode like a beanbag with the first bit of pressure. Instead of an exoskeleton, like that of the wasp or the sausage, we have an endoskeleton. Inside our bodies is a hat rack of bones on which all our organs can be hung. Flay a man and the cohesion remains unharmed; a casing isn't necessary. In this respect, a human being is the opposite of a sausage. A sausage cannot do without its casing because all the bones have been removed from the filling. Bones in your sausage are bad for your teeth. But bones in your turd would do considerable damage as well, especially to your delicate intestinal wall. So instead of bones, a wise turd contains fibre to serve as its skeleton—something it cribbed from the plants. Plants also manage quite nicely with fibre instead of bones. Intestines produce good, sturdy turds from the plant fibre in food. You eat not only to please your taste buds but to please your backside as well, and both ends make demands that comply with the Biblical injunction 'Render unto Caesar that which is Caesar's and to God that which is God's'. Unfortunately, the commandment seems to have lost some of its force. Belly and chest get their share—and often more than that—of well-cooked, refined, fibre-free edibles, but gut and anus can't count on any fibre from a fast-food diet.

There isn't a bit of straw left in the turd for the shit to hold onto, and it has no choice but to slide out the backdoor.

For an example of how beautiful an old-fashioned fibre-turd can be, take the horse. How glorious it must have been to be a coachman back in the day and to watch from the box as the noble animal lifted its tail, brought forth great masses of excrement neatly bundled into little bales, and then, without any wiping, piously shut its equine arsehole like a virginal oyster. Thanks to the horse's body, which can manage quite nicely with a portion of its feed, there's more than enough building material left over to produce a pile of manure. An elephant goes one better. Despite 25 metres of small intestine, 1.5 metres of appendix, 6.5 metres of large intestine and 4 metres of rectum, half the elephant's food goes undigested. Its droppings know a trick that your turds don't: they float. Turds tend to sink as a rule because, like all other organic tissue, they're heavier than an equivalent volume of water. But when fibres are excreted their bacteria keep emitting gas, which makes them act like built-in swimming tubes. The strict vegetarians among us also have turds that float, although they only notice it when they use a toilet that contains plenty of water. An extra large quantity of bacteria advance on the undigested fibre in their rectums, giving their excrement the light yet cohesive quality of a particularly successful doughnut. In certain parts of Africa where the human diet is heavy in fibre, the people shit up to four times as much as we do in the form of lovely, upside-down toadstools, which float along gracefully and impart a cheerful look to a village built on the banks of a slow-moving river.

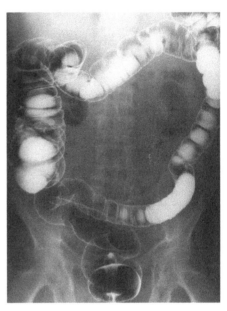

X-ray of the large intestine. The contents have been made visible with the use of barium.

A turd's cohesion determines its length. The softer the shit, the shorter the turd; in the case of diarrhoea length isn't even an issue. You could try to produce a really long specimen if the intestinal content were good, but you'd have to have a well-developed pushing technique and lots of patience. It would be more comfortable to divide a single portion into several turds, like a butcher getting a whole string of sausage links out of one long intestine. It's all a matter of pinching. Sometimes, however, you have no choice, and the shit decides on its own whether it's going to be a bowlful of sludge or a pile of pellets. For other species of animals that choice is permanent. Instead of one continuous turd, a rabbit can shit up to five hundred pellets a day. Deer do the same, but in their case the pellets are bunched together like gigantic brown raspberries. You can tell how much an insect shits by studying caterpillars that live together in one big cocoon. Once the caterpillars have turned into butterflies and flown away, their black droppings remain in the cocoon. In a small woodland, all the caterpillars together easily shit half a tonne per hectare. The little piles of sawdust that are shat out by woodworms are nothing in comparison. But those who fail to heed this warning will eventually find themselves minus a roof beam—or a cathedral.

You can tell how much a person shits by weighing it. But how do you weigh your own excrement? There are two methods: a yucky one and a tidy one. For the first method you simply shit on a set of scales. A handier way is to weigh the shit while it's still neatly tucked inside your body. Then go to the toilet and weigh yourself again when you're finished. The difference is the weight you're looking for, and it's bound to be gratifying: easily half a pound, a block of butter. But sometimes all you get is about a hundred grams. An average of 175 grams is fine.

For the sake of convenience, let's say the specific gravity of shit is equal to that of water—one kilogram per litre. So an average amount has a volume of 175 mL. At a diameter of 3.5 centimetres, you end up with a length of 18 centimetres. In terms of daily bowel movements that's 125 centimetres a week, 65 metres a year, 5 kilometres a lifetime. If you shit 175 grams a day, that's 65 kilos a year. But even the best shit machine breaks down every now and then, and there are days that you don't excrete at all. A good rule of thumb is 50 kilos. In order to produce that in one year you would have to eat a thousand kilos. That's 1:20 for a human being. On average. The variation is enormous. Sometimes all you produce is a mouse dropping, and sometimes the toilet bowl is so full you think you can actually feel the top of the pile brushing against your arse. The size of the biggest human turd will always be unknown as long as no contests are held. Boys try to see who pisses furthest, not who shits the most. We know what you'd have to do to win such a shitting contest. You'd have to increase the water content. You'd have to pump up your turd with water like a battery chicken. But then there's the danger of diluting it and causing diarrhoea. If you take a good solid turd with a 20 per cent dry matter content and water it down to 10 per cent, you'll have shit coming out of your ears like bouillon. You have to bind the water, with fibre. In order to produce a winning turd, the idea isn't to drink extra water—a healthy body has enough water on board—but to eat extra fibre. The fibres swell with water and your turd becomes both thicker and heavier. You shit more, even though you eat less.

This brings you to a revolutionary idea: can you shit yourself thin? Is there anything like a shit diet? It sounds attractive. Ordinary dieting is no good. It only makes you hungry. Hunger makes you eat, and eating makes you fat. Why not do the opposite: don't put in less but take out more? If you defecate four times a day instead of just once, you'll lose four times 175 grams = 700 grams, which is 5 kilos a week and 20 kilos a month! Within a year you'd have shat yourself from the face of the earth!

A living advertisement for the shit diet is the Dutch writer Maarten 't Hart, skinny as a rail all his life and an inexhaustible shitter:

> When I was in nursery school no one wanted to sit next to me on the bench because, according to my classmates, I 'let such awful farts'. Sadder still, I had to keep going to the toilet to do what they called a number two. But at first Miss Dekwaaisteniet refused to believe that after going out for my first number two I could feel a second one coming on only half an hour later. 'You've already had your time on the pot!' she would shout incredulously. So I'd try my best to hold it, but usually that didn't work and I'd end up 'doing it in my pants', as my mother put it so expressively, and sure enough,

a terrific stench would make its way through the Dam School. I can see it all before me as clear as day, my classmates leaving off from their cutting, pasting, and colouring activities and pointing their accusing fingers in my direction. It wasn't unusual for the girls to burst into tears.

And that's the way it's always been. 'Having to do a number two six or seven times a day probably accounts for the fact that I've always weighed 76 kilos, even though I can dig into my meals with both fists,' Maarten 't Hart writes with satisfaction almost sixty years later in *The On-Deaf-Ears Diet*. 'I can eat anything at all, as long as it makes me shit.' 't Hart is naturally gifted, but it shouldn't be that difficult for an ordinary person to go to the loo more frequently. There are pills for this purpose—or stew, or coffee. But you can't keep it up very long. One look behind you and you'll understand the reason why. A runny mess. You'd be losing water instead of calories. There's scarcely any extra shit to be got rid of from a single meal. At the very most you dilute it, like orangeade with water or whisky with soda.

Your intestines are just too smart for you. Laxatives put them into overdrive, so they don't have time to absorb the nutrients that make you fat. But they also don't have time to reabsorb a sufficient amount of water, so you end up with diarrhoea. The only thing left is to get older. In old people, the papillae in the intestines that absorb food get worn out, so that more food is transported unused. That's why old people are often so thin. But you also have people whose intestines work too well, which makes them fat. They gain weight 'just by breathing'. That's not necessarily their own fault. It often has to do with their bacteria. Some of these bacteria convert the waste product (CO_2) from other types of bacteria into human food, which is then absorbed in the large intestine—handy in times of famine but disastrous in our age of milk and honey, if you're already on the heavy side.

Those who do see something in the poo diet are guilty of misjudging their intestinal tract. Its job isn't so much to get rid of things as to absorb them. When the intestines are working well, they absorb so many nutrients that there's hardly anything left to excrete. The excretory organ par excellence is the not the intestine but the kidney. It gets rid of really superfluous and harmful products like table salt and protein waste. You hardly lose any weight at all through your kidneys. What you pee is mostly water. A pee diet is as senseless as a shit diet. If you were to try to pee a kilo off your body weight, the once proud and mighty stream would soon dwindle into increasingly scanty drips until even the hardest squeezing wouldn't help. The salt in your body jealously retains fluids. Your kidneys are too smart for you, too. So is our body always out to get us? Who knows? But you can always fool it—with fibre again. Whatever food may be found in fibre is almost impossible for the intestines to get at, but they do get tricked into thinking they've had a good meal. As long as the stomach and intestines are full, you don't get hungry, and both the intestines and their fat boss are satisfied. The best way to lose weight is by eating hay. It comes out no different than it was when it went in, unopened return mail, like a nun in heaven. But not even the worst glutton has a hankering for hay. That's why we give hay to the cows, who know what to do with it better than we do. But we turn the cows into roast beef, which our intestines do have a taste for. Your best bet is to have a roast beef sandwich on wholewheat bread: the roast beef for your taste buds, the full grain for your bum.

A well-filled stomach makes for good shitting. If the filling turns out well, you can work it into any shape, as every butcher, every baker and every biologist knows. Shape is what a good turd has over such amorphous secretions as snot or ear wax. And the best shape for a turd is the worm shape, which it naturally assumes from the intestines that produce it. Intestine and turd fit together like a biscuit in its tin, an underground in its tube, a mole in its tunnel. We can thank our lucky stars that our turds don't clear a path to the exit in the shape of a cube or a garden rake.

Like a train, a worm-shaped turd, no matter how long, needs no more than a narrow tunnel. If the first carriage can get through, the rest of the carriages will follow. Both carriage and turd keep their shape throughout the ride. The last section of intestine, the rectum, is as straight as its name—Latin for 'straight'—suggests. It's only when they nestle into the toilet bowl that they assume the elegant spiral shape of a cartoon turd. With its pointed tail standing straight up, the turd looks as if it had been pulled through a ring. And in fact it *has* been pulled through a ring. The anus.

It can be tempting to see your own turd as a living animal. For those who gaze with tenderness on their offspring, nothing looks more like a litter of sleeping puppies than four turds, intimately curled up together.

What a turd lacks in order to be numbered among nature's living creatures, besides four legs, is reproduction. There are no male and female turds. Never has anyone ever seen turds clasped in a loving embrace by the light of a full moon in an effort to make little droppings. Turds can divide, but the halves don't grow to reach the original dimensions. The fact that there are turds at all is only because their manufacturers themselves manage to reproduce.

And yet a turd lives. It's bursting with life, like a reef full of fish or a rainforest full of monkeys. In every gram of shit there are more than a billion living bacteria. Together with their deceased family members they form half the poo's volume. If you could hear the life in your turd, you'd be listening to the deafening noise rising from a hundred billion little throats. The reproduction that the turd itself does not exhibit is more than made up for by that of its residents. As long as the monster stays warm—watch it steam!—they'll divide a couple of times an hour.

You'd hardly dare wipe your backside if you knew how much it was teeming with bacilli. But then you wouldn't dare eat any sausage, either. A sausage is also crawling with microorganisms, externally and internally, not ones that accidently dropped in but ones that were deliberately planted there. The butcher adds lactic bacteria to the sausage to attain the right acidity and staphylococcus bacteria (known from boils) for flavour. That white layer on the sausage is exactly what you were afraid it was: mould. Bacteria, moulds and other microorganisms have already started eating. Biologists call it rotting, but butchers prefer to speak of ripening. It's good work. These microorganisms make sausage taste better, and they make the gut healthier. Gut bacteria are the little gnomes at work in your tummy. They diligently clear away poison, knock together vitamins and dig new mines in yet undigested fibre.

In addition to *Escherichia coli*, the ultimate in intestinal bacteria, there are hundreds of other varieties that usually have better things to do than make you sick. The most useful thing about these good bacteria is just that they're there. The presence of a good bacterium means the absence of a villain. But evil always finds a way, especially if it streams in in torrents. That's how holidays get ruined. The *E. coli* bacteria living in the water in holiday resorts abroad have a slightly different appearance, reason enough for our xenophobic intestines to erupt into cramps. So instead of seeing the sights you make repeated visits to the toilet, or what passes for a toilet in that particular country. A third of all tourists spend their first week in the loo. Only after several rounds of flushing does your intestine recognise the weird strangers as family, and peace is restored. Couldn't our intestines have behaved more hospitably from the start? Perhaps, but people who have overly hospitable intestines usually succumb to an infectious illness sooner or later.

Of course it's important to fight pernicious bacteria. But excessive hygiene is counterproductive. You end up killing the robins and the sparrows of your own ecosystem along with the rats. And a total extermination of all the mischief-makers isn't possible anyway. Better to strive for a good balance by leaving it up to the lodgers themselves. They have more knowledge of this than you do. They live in a balance that was achieved during your first years of life, a balance so stable that experts can divide the intestinal flora into types in exactly the same way that blood is divided into blood groups. There are three, each characterised by the bacteria *Bacteroides*, *Prevotella* and *Ruminococcus*. Which group you belong to partly depends on what you eat. *Prevotella*, for instance, feel most comfortable in the gut of a sugar-lover.

Nothing tops a stable intestinal flora that keeps out the foreigners. But what if the established order itself is a mafia state? No use swallowing a dose of good bacteria then; they don't stand a chance. The only thing that can help is an actual transplant. You might need a transplant if a heavy course of antibiotics has wiped out your original intestinal flora, the good along with the bad, allowing a tenacious colony of clostridia to move in—with all the resulting diarrhoea and even death threats. In a transplant, a hefty portion of faeces from a healthy donor is introduced from above, through the nose, or from below, through the anus. Doctors are confident about the future of faecal transplantation. First of all, there are fewer people in need of someone else's shit than those in need of someone else's heart or liver; and, second, it's much easier to find a donor than in the case of a heart or liver; and, third, it really helps.

Bacteria are too small to be seen darting around in your shit, licking here, peeing there. Maybe it's just as well. It's all the more terrifying to actually see something white wriggling through all the brown. Optical illusion? No. Pinworms (*Enterobius vermicularis*). Just looking at them makes you itch, if you haven't had them already. At night the adult females creep through your rectum to your anus to lay their eggs. That itches. Itching makes you scratch, and scratching makes you itch. And you should be glad it isn't roundworms (*Ascaris lumbricoides*). At thirty centimetres in length they're at least fifty times bigger than pinworms—really giant earthworms. Adult females lay some 200,000 microscopic eggs a day in your shit. The larvae in their golden-brown eggshells wait in a damp corner of the big bad outside world until they're picked up and ingested by a less hygienically inclined human. That marks the beginning of a long, adventurous journey through the human body, in which the larva hopes to be coughed out of the lungs by any means necessary and immediately swallowed down the same throat before settling in the small intestine as an adult roundworm, to fleck your faeces with eggs. Even though you can't see them.

What you might come across are little white bits in your shit. If you do, you have tapeworm (*Taenia solium*). What you see are the worm segments, each one packed with fertilised eggs. At ten segments a day, each one good for 100,000 eggs, you easily arrive at a million. The worm won't miss them because it has thousands already and makes them in its neck as quickly as it releases them from behind. The tapeworm doesn't need a mouth or intestine to eat; it feasts with its entire body surface on the half-digested manna around it. Supposedly this makes you lose weight. At one time it was the fashion in Hollywood to treat yourself to a tapeworm. It wasn't expensive. You simply ate a raw or undercooked piece of intermediate host: ground beef with pig tapeworm for the Christians, steak tartare with beef tapeworm for the Jews, and a lick from a dog full of canine tapeworm for the vegetarians. People did indeed lose weight, but that was mainly from the diseases they contracted. Tapeworms secrete harmful substances and can block the intestinal tract with their bodies. If you remained healthy with such a worm inside you, at most you'd end up eating more, just like pregnant women eating for two.

As notorious parasites, intestinal worms have long been regarded as degenerate. While cautious animal species living in the open air climbed up the evolutionary ladder step by step, the worms in our intestines neglected both their behaviour and their entire anatomy because they didn't need them for the cushy existence they were living. Blind, deaf and toothless, they just hung around in the tepid warmth of the gut. Actually, if you're going to be a burglar and a freeloader you've got to be very cunning. Compared with a predator like the wolf or the lion, parasites are extremely sophisticated when it comes to putting food on the table.

Running stupidly after their food, aggressively swinging their claws, murdering and terrorising are not their approach. Instead of killing the goose that lays the golden eggs, both parasite and host spend their whole lives eating together. But in order to do that the parasite has to find a way to get inside the host, often in the guise of an unrecognisable stage in its life cycle, and then to stay put, despite safety precautions like an elaborate immune system. How does such a flawed animal manage to survive in someone else's body despite its increasingly striking dimensions? That's something you ought to know very well. After all, you did it yourself, years ago, for nine long months.

Checking for worms is a good reason to take a look behind you before you flush. If anything's wrong you should see your doctor, and take your faeces with you. It will involve a lot of fumbling with pots and spoons. How do biologists manage with animal shit? They just scoop it up, often with their bare hands. Most animal turds are remarkably dry and clean. Those made by predators are sometimes like perfect cigars. Usually there's a pointed tip where the cigar and its maker parted company; in the case of the fox there's a wisp of fur furnished with a festive tuft.

Pellets, turds and pats are the turds produced by herbivores as a rule. A cursory inspection confirms it. The fibre turns the whole mass golden brown, and sometimes it even contains cheerful orange, red or purple berries, which accounts for the not unattractive sweet scent. Pellets come in many different sizes. Pellet no. 1, with a 5-millimetre diameter, is made by squirrels; no. 2 (10 millimetres) will have been the work of a rabbit, since no. 3, 20 millimetres across, comes from hares. Rabbit pellets lie together on raised areas.

The turds of insectivores do credit to their name. They're full of wing cases, legs and other parts of the prey. A good example are the turds of hedgehogs. There are very fine insect bits in the excrement of bats, often piled up under the places where they sleep. Many mammals have anal glands near their anus which give each turd an individual stamp, like the production number on a cheese or a radio. That message is not meant for human noses. Yet we can sometimes tell what kind of animal produced the turd by its smell. You don't have to look to identify the shit of fish eaters, for example.

The hardest ones are the birds. The formlessness of their faeces gives us little to work with. Yet there is a certain structure in their shit and pee combination. Often the shit is like a dark yolk in a runny fried egg, sitting in the midst of the egg-white urine. To identify the species you can look it up in *What Bird Did That?* by Peter Hansard and Burton Silver. In addition to photos, the authors provide detailed descriptions of excrement based on a main division of splerds, sklops, sploods, schplutzen, and schplerters. A sklop, for example, has a 'small, clearly defined envelope and nucleus of roughly equal proportions. No tendency to lobe.' Schplerters, on the other hand, are 'large [with] multiple extended and detached lobes', and they're splash-shaped. The best place to identify these types is on the windscreen of your car, which offers a good, smooth surface. A more comfortable place for bird-watching doesn't exist. Nor can it get any simpler. After a bit of practice you'll be able to identify 'an extended sklop with an attractive cloud-like appearance' as the 'moist, loose, fragile' splotch of the blackbird, certainly if the grains that are its regular feature are not located in the nucleus but in the envelope. In season, the rather loosely formed schplutz of the song thrush can mainly be identified by the 'bright streaks and flecks' that berries and other fruits have left behind in the creamy goo. All this comparison with white sludge (and its occasional crunchy bits) puts you in mind of popular dessert puddings. In his book *Merde*, leading scatologist Ralph Lewin doubts the accuracy of this approach:

> Although this booklet [*What Bird Did That?*] is certainly among the most entertaining of all the works cited here, one is led to question the authenticity of the information in contains. (Healthy

geese, I know, do not make splattery splotches like those they have illustrated, and many other entries, though perhaps more plausible, are equally suspect.) However, as an amusing item for coprophile-bibliophiles the book is to be strongly recommended.

Personally my suspicion was mainly aroused by the introductory pages and the material in the back of the book, where literature like A. Crichton's *The Chinese Hat Method of Splay Collection* is included, and successors to *What Bird Did That?* are announced, such as *What Toad Did That?*, *What Camel Did That?* and *How Do You Know It Wasn't a Dromedary?*

As so often happens, nothing beats observation and research. The problem in this case is the short lifespan of the material. Faeces are made to decay as soon as possible, otherwise the world would be a much less attractive place to live in. If they don't get eaten, they shrivel up, dry out and discolour, thereby losing their value for the researcher. Serious collectors have to preserve their material. In order to write his *Field Guide to Animal Tracks*, Olaus Murie collected more than 1200 specimens, dried them, varnished them, stuffed them and labelled them. For her collection, Annemarie van Diepenbeek made use of modern kitchen appliances:

> Faeces have to be thoroughly dried, preferably followed by several weeks in the freezer in order to kill off any insects and microorganisms. The drying process can be accelerated by placing the specimen in a microwave oven at a low setting. Harmful organisms can be killed off by reheating at a high setting. Objects containing metal, such as pellets with bird rings, should not be heated in the microwave.

Occasionally we find that nature has already done the collector's work, with faeces that have been petrified into fossilised turds: coprolites. One of the first collectors was Dean Buckland, who identified the stones he found in a cave along the coast of Dorset in 1823 as hyena droppings. As becomes an eccentric Englishman (Buckland also studied the footprints of people with wooden legs and liked to serve his guests roasted mouse or crocodile), he was soon showing off a whole table full of petrified turds. The high point of his collection, of course, was the turd of a dinosaur. These turds tell us a great deal about what the animal ingested. The largest specimen so far is a 45-centimetre coprolite that was found in Canada in 1995 and came from a *Tyrannosaurus rex*. In it were the remains of a young herbivore the size of a cow. Collectors pay thousands of euros for a dinosaur turd. That sounds like a lot, but it must give you an extraordinary feeling to stand there with a giant turd in your hands shat by such an animal from its enormous arse millions of years ago.

More recent but equally interesting are the coprolites from our own species. The value of the most expensive human shit in the world has been estimated at 20,000 British pounds. It's the Lloyds Bank Turd found under the building of the same name, 1000 years old but in mint condition. Usually an antique dropping like that is first soaked before being spun in a centrifuge, then sifted, and then further analysed. This procedure has brought to light the eggs of roundworms from specimens more than 24,000 years old. If there's still DNA in it even the sex of the defecater can be identified. The finest investigation was carried out by Andrew Jones of the University of York. He kept eating different things until he was able to shit out a turd that was exactly like the one whose contents he wanted to discover. He had to eat a whole lot of fibre to work up a genuine Viking turd, but that didn't stop him. They still exist, eccentric Englishmen.

6

What a Relief

You're older than you think you are—three-quarters of a year older, to be exact. Your life began at conception. Even as a single cell you were completely yourself, with a unique genetic code, all your talents and bad manners already pre-programmed. And your neighbour was completely himself, as was your auntie purely herself. Long before you were delivered, a number of far-reaching alterations took place. First you looked like a paramecium, then like a tadpole, and later like a monkey. A tail came and went. Submerged in the amniotic fluid, you used your mother as a snorkel, embryonic brains filling themselves with embryonic thoughts. You experienced more in your first months than in the rest of your whole life; by the time you were born the most exciting things were already behind you.

But that doesn't count. You've only been a member of humanity since your birth, since taking leave of your mother, coming into the world. Playing outdoors. Suddenly there you were. A miracle. To celebrate the festive opening of your life, your parents sent out cards. To say you were there, and what they thought your name should be. There was laughing and drinking, as there is at the beginning of anything new: the completion of a house, the presentation of a book, the launching of a rocket or a magazine, the incarnation of a god.

For nine months you lay like a turd in your mother's body before being ejected, just like every other piece of excrement. But you no longer remember that. You missed the most important moment of your life. Nothing for it but to pay better attention later on, when the next most exciting moment comes: death. At least you can imagine what your beginning was like, though: the increasingly cramped feeling there inside, the growing realisation that you had to get out before it was too late, the world around you suddenly squeezing and squeezing, the light at the end of the tunnel getting brighter until it made your eyes hurt. It was time to re-route your circulatory system to your lungs; your navel was disconnected and there you were, taking your first breath. No space travel or heroin can hold a candle to it.

You have no memory of this, but your mother's memory more than makes up for it. While you were getting born she was giving birth. For her, too, the tension mounted, the fear of being torn apart, the necessity for drastic measures, followed by the catharsis: the delivery. It's a secret women's experience from which men are thought to be excluded. But men are not completely unfamiliar with delivery. Men also cherish the fruit of their innards. Men shit, too.

Shitting is childbirth on a small scale. In both cases, there's something inside that has to be let out, before it's too late. Given the fact that a friendly appeal to come out meets with deaf ears, the only choice is to resort to violence. Here an awkward construction has its revenge. It would have been so much better if our fruit had grown outside our bodies rather than inside. This isn't as strange as it seems. It's how trees do it.

Trees are smarter than people. Dangling from branches, their apples and pears grow freely in size until they've had enough, then they lazily drop to the ground. Thanks to this system, one girl tree can bring forth hundreds of fruits at once; a human girl complains if she has to bear twins. Dead leaves, old stamens and other waste simply yield to the force of gravity, just like the fruit. Birds also raise their embryos outside the body, sitting on their eggs. Rubbish is tossed out over the edge of the nest, and the young fly away when they're ready to go. The worst place for a nest is in your belly. For something so stupid you'd have to be a mammal.

Of all the mammals, the human being is the most troublesome to bring into the world. Its head is too big. Try putting on a turtleneck sweater. You can barely get it over your head. So you can imagine what a bother it must have been to push that head through your mother's cervix when you were born. Because of the disparity between the baby's head and the mother's pelvis, a human birth most resembles an attempt to push an apple through a funnel without making applesauce out of it. The apple brings it off with striking frequency, but the funnel sustains serious injuries. Sometime something tears, the bladder implodes, the pelvis collapses.

Not so very long ago, three out of every hundred Dutch women never left their childbed. But even if all goes well, the expulsion of a baby is a huge hassle. That huge hassle is what we call birth, although the term 'delivery' more aptly describes the situation. Sometimes even shitting can be another childbirth. Ironically enough, this is mainly true of the mother's first bowel movement after the birth of her child, when the stomach muscles slacken and the rectum is still half unconscious from all the neighbouring violence. Usually, however, shitting a turd is a piece of cake compared with the delivery of a baby. Most of the time it's just plain delightful, a childish pleasure, like messing about with tomato ketchup or writing your name with syrup. The pleasure is more in the production than in the product. You see that frequently with creative hobbies. A loom, with all its shafts and levers, is a great deal more beautiful than many of the lengths of cloth that artists make on it, and if I didn't find honey so distasteful I would have started beekeeping long ago, with all that interesting traditional fuss and bother that goes into it. If shitting is a versatile activity that you can keep on refining all your life, the product, no matter how successful, is just something to be tossed away. Still warm from your body, it's cruelly disowned like an illegitimate child. A legitimate child, on the other hand—a child you bore yourself—is something to love. In that case, the product is considerably nicer than the production. There doesn't seem to be much you can do about it, since God drove Eve out of Paradise with the words, 'I will greatly multiply thy sorrow and thy conception; in sorrow thou shalt bring forth children.'

With our big heads and our narrow birth canals, the Creator has been generous in His contribution to the sorrow of childbirth. What makes the whole thing even worse is the absence of cooperation on the part of the child. Excrement lets itself be driven out passively. By growing, both child and turd give a clear sign that they need to evacuate, but the actual ejection falls entirely on the body. In the absence of block and tackle, hook and eye, or wheel and axle, it's all a matter of pushing. And that depends on muscle power. The contents are pushed from the uterus or the rectum like toothpaste from a tube. The fingers needed for squeezing are built into the walls. At the same time, muscles in the abdominal wall and the diaphragm put the abdominal cavity under pressure. To underscore the similarity between the mechanics of childbirth and defecation, it might not be a bad idea to speak of labour when referring to defecation. Biologists use the term 'giant migrating contractions', when the sphincter muscles squeeze the large intestine almost completely shut in some places. These contractions shift along a great distance, forcefully pushing the contents towards the exit. This happens only about six times a day, usually shortly after rising, after meals, and after coffee or tea. The sphincters squeeze during the rest of the day as well, but these haustral contractions aren't nearly as deep or as fast. Their purpose is not to propel the contents forward but to mix and knead it, so that all the bits of pulp come in close contact with the wall in order to exchange substances. Giant migrating contractions move at a speed of one centimetre per second. When enough shit is ready, it's stored at the end of the large intestine. The rectum itself is empty for the most of the day. When it is full it swells up, which comes to the attention of the stretch receptors in the rectal wall. If there are 60 to 100 millilitres of faeces in the rectum, the receptors send a message to the nervous system: you feel the need to go. Sometimes you feel it ten minutes after you've eaten, when your food couldn't possibly have reached your rectum yet. That means your full stomach has made a preliminary announcement with the help of hormones such as gastrin. You can answer the call of nature immediately, but if you're in a meeting or stuck in rush hour traffic it might be wiser to hold off for a little while. When that happens the turd usually slips back into its limbo, the S-curve between the large intestine and the rectum, the sigmoid. But don't get carried away. If the amount of shit in the rectum rises to more than 250 to 400 millilitres, the dikes will burst, whether you like it or not.

Timing your defecation is a matter of cooperation between your head and your belly. As long as the dikes can easily hold back the tide, the head can exercise veto rights—to a certain extent. It can say *nyet* to the general consensus. Likewise, there's no way you can defecate if your belly doesn't feel like it. Shitting follows an internal rhythm, like breathing, sleeping and menstruating. A healthy belly likes to do it once a day, more or less synchronous with other bodily functions. But even if you shit twice a day, or twice a week, you can still live to be a hundred. John Harvey Kellogg thought three to four times a day was a minimum requirement for staying healthy. The only way normal mortals could achieve something like that was by irrigating the intestines between bowel movements. Whether this practice could assure you of a hundred-year lifespan is doubtful. On the other hand, according to Dan Sabbath and Mandel Hall in *End Product*, the man who shat only on Saturday evenings, and then never more than every other week, lived to be seventy-seven. He did that for nineteen years. But there are limits. After being held in for three weeks, the shit gets so compacted that it can only be prised out with the help of an instrument or an operation. The record for not shitting, by the way, is held by Adam and Eve in Paradise. According to the biblical scholars, they didn't have to defecate because their food was immaculate. There are others who disagree. They argue that Adam and Eve were created with full intestines, just as they were made with navels, because the whole world was created with a prehistory. Long before this question arose among theologians, the Essenes, a Jewish sect, may very well have imitated Adam and Eve. The Essenes lived in model communes where everything was held in common and the love of God was regarded as the greatest wealth. According to the Roman philosopher Porphyry, 'they had such simple, sober eating habits that they did not have to defecate until the seventh day after partaking of their food, and they spent that day singing hymns to God.' Just picture it, the sect members, their thighs clamped together for days, their heads getting redder and redder from grim-faced fanaticism.

In modern times, the place to look for faecal suppression is the animal kingdom, although you might not imagine it with all those dogs around. The sloth is laziest in this respect. This South American creature sleeps fifteen hours a day, but to suggest that it therefore must be awake for the nine remaining hours is overdoing it. A sloth is never really awake. Usually it just hangs around from branches, upside down, and if it moves at all it does it in slow motion. Other than that, it has hardly any habits; a sloth just doesn't have them. Except for one: once a week it comes down to the ground from the leafy canopy to defecate and urinate at the foot of its tree. In this way, according to foresters, it keeps from fouling itself. Others say it does this to fertilise its tree, or to prevent the sound of shit falling on the leaves from revealing its location. In any case, it doesn't have to go more frequently on account of its slow digestion. Something similar happens with hibernating bears. Thanks to a good solid plug of fur and shit, their anuses are shut for months and they don't have to get out of bed to defecate or urinate. Human beings also have a splendid biological mechanism for this, which wasn't discovered until 2012. At night your bladder is simply able to hold more urine. There's less of the protein connexin-43 in the bladder wall, which makes the bladder wall relax. In addition, the kidneys make less urine at night. Unfortunately, this mechanism starts faltering as a person ages, so old people end up breaking their hips at night while stumbling to the loo.

Even the best device fails when the fuses blow. When you're really terrified you pee in your pants. Shortly before a school exam, when an explosion occurs nearby, or after a knock on the door that might mean your death, staying dry is a tremendous struggle. The brains in your intestines, egged on by hormones like histamine and prostaglandin, push aside the brains in the head and take full charge. It's as if your body had risen up in revolt against your mind. There are infamous stories of the soldiers in the trenches during the First World War who stood with their feet in the mud and their legs in their own shit, which did not promote fighting efficiency. Yet that was exactly what the body intended. In the face of great danger, hormones and nerves shift the body over to a state of war. All efforts are then directed towards either fight or flight. Organs that have nothing to contribute to the cause, such as the intestines, are temporarily cut off from blood and other provisions. This happens to many game. An antelope shits its brains out at the sight of an oncoming lion. It quite literally drops its ballast. An antelope turd may not weigh much, but it can mean the difference between life and death if the animal can run just a little faster without it. For birds, dropping ballast when danger approaches is even more advantageous. Ralph Lewin knew an ecologist in Malaysia who studied the internal parasites of birds. In order to collect them, all he had to do was put the host birds in a plastic bag for a short period of time and the parasites would appear in a wave of anxiety excrement. But the prize for the most excellent mechanism goes to the caterpillars of the skipper butterfly. When danger threatens, they shit with such force that it makes their attacker run away. Their poo shoots a metre and a half into the air. If we could do that, relative to our body length, we'd be shitting a distance of sixty metres.

Shitting yourself from fear has a sad parallel in the bearing of a child. While the English soldiers were befouling themselves during the liberation of the European continent in World War II, their wives in London were running a greater risk of having miscarriages during the V1 and V2 bombings. In both cases, stress hormones and traumas disrupted their bodies' normal operation.

'Normal' defecation is something most people don't know anything about due to a lack of examples. Occasionally you get to see the birth of a child on TV, or photos that make the rounds during a visit to a new mother. But images of a session of shitting would chase viewers or visitors away. Even when you're alone it's hard to actually see how shitting works. That accounts for all the fussing and fumbling. The first lesson at every trade school was: always watch what you're doing. Keep an eye on the work at hand and make sure you've assumed the right position. To start off, watch how the animals do it. Many of them look as if they were giving birth when they're shitting. They squat with their tail in the air. This is the easiest way for a turd or a young to slip out. You can tell from its physical attitude that your cat is about to take a shit, which is intended for that one purpose alone. Dan Sabbath and Mandel Hall refer to two German researchers who inserted electrodes into the brains of cats. When they stimulated the septum pellucidum and then the stria terminalis, the cats assumed the proper position and defecated. 'This is very Prussian. Unfortunately, when the shocks were given in reverse, the cats defecated, *then* assumed the defecatory posture.'

Human beings, with their upright gait, are cut out for squatting. Primitive peoples squatted at every opportunity, and many Asian people still do. In the West, people revert to the four-footed stage of evolution. They sit on furniture with four legs all day long, and when they go outside they move around in a chair on four wheels, their car. This way of sitting takes its revenge when they relieve themselves of a child or a turd. Giving birth on a chair, as was customary in earlier times, is a better position than giving birth in a bed, but it's still more awkward than giving birth while squatting. Shitting in bed is not something you do voluntarily; it's generally seen as an uncomfortable experience. But even the most ordinary toilet gets in the way of the defecation process. When you sit, passing a turd is hampered by the puborectalis muscle. When relaxed, this muscle causes a twist to form in the rectum, thereby dividing it into three compartments. The partitions bear the weight of the faecal column in the intestine in order to relieve the anal sphincter. So you can easily remain housebroken in a standing or seated position. A human being squats naturally in order to shit. In this position the levator ani muscle automatically swings into action: the twist in the intestines slackens, the buttocks spread apart, and the contents easily slide out. In Europe the French have known this for a long time. They retained the squat toilet, so feared by tourists, with its two footsteps on either side of the drainage hole. As a tourist this is where you really felt like a foreigner because of the fear of falling over; you were never sure whether the door handle you were clinging to would hold. But it's all a question of practice. Fanatical proponents of squat defecation even manage to shit by squatting on the upper edge of an ordinary toilet (with the seat raised). They insist that shitting while sitting leads to constipation, haemorrhoids, appendicitis, incontinence, irritable bowel syndrome, enlarged prostate and intestinal cancer. Less fanatic followers place a foot stool in front of their ordinary toilet bowl to imitate a squatting posture. For the beginner, it's enough to place the newspaper (which you wanted to read anyway) on the floor so you have to bend way over in order to fulfil two needs satisfactorily at the same time. Sitting straight up on the toilet and reading the newspaper is an outgrowth of our movement-impoverished culture. Instead of football and running, the first thing our children should be taught in their school gymnastics classes is how to squat. Then they'd be able to relieve themselves anywhere, without a chair or a toilet bowl. It's a mystery why something as unhealthy as defecating sitting down has become the norm in our health-advice-infested world, but one explanation is unavoidable: in our zeal to get something as filthy as a turd into the toilet as fast as possible, we've forgotten how to get the same turd out of our bodies with the same speed. But you're never too old to learn. Lower your arse as far as you can, with your legs spread wide and your arms wrapped around your knees for balance. In this way your thighs automatically press against your lower body to help push the shit out. Good luck!

German soldiers even had rules for shitting during the First World War: not from the side but in the 'squatting position'.

Having reached the rectum after its long journey, the turd is ready and wants to get out. But it's not quite that easy. Usually it comes up against a closed door: the anus. And before it reaches the back door itself, the shit has to pass through the gateway to the anal canal, where the intestine bores through the muscular pelvic floor. The twist in the puborectalis muscle serves as a storm door. When it pulls the intestine forward it forms a sharp 80-degree angle. Should any shit slip through anyway, it gets sent back like a truant child. This is where the anal canal makes use of a good trick: it squeezes together at the exit more often than at the entrance. The back door itself shuts as firmly as the cap of a tube, except that instead of a twist-off cap there's a sphincter muscle. No turd, no drip, no cloud of gas can escape. 'Poems are made by fools like me,' urologist Victor Marshall taught his students in order to impart a bit of respect for the anus, 'but only God can make a sphincter.' The anal sphincter is made with a double wall. The inner wall is just the extension of the intestinal sphincters. Like every smooth muscle, the action of the innermost sphincter is involuntary. It simply relaxes as soon as a sufficient amount of shit presents itself. But most of the time it keeps itself tightly closed, a condition of which it never tires. The outermost sphincter, on the other hand, is dead beat after a minute of squeezing. Like all true striated muscles its action is voluntary, although it's sometimes disobedient.

Once your brain has granted permission, the puborectalis muscle relaxes, the twist in the rectum slackens to a 125-degree angle, and the pelvis drops, anus and all. Now the pelvic floor is like a hungry funnel, ready to receive the turd and push it out with the help of the diaphragm muscles and the muscles of the abdominal wall. But sometimes something gets in the way. Oncoming traffic. That's what happened to a carpenter's apprentice from Great Yarmouth in around 1725. According to an article in *Philosophical Transactions* with the telling title 'An Account of a Fork Put Up the Anus, that Was Afterwards Drawn Out Through the Buttock', he reported to Mr John Ranby, Surgeon F.R.S., with the tines of a fork protruding from his bottom.

Being costive, he put the said Fork up his Fundament, thinking by that Means to help himself, but unfortunately it slipt up so far, that he could not recover it again. It is 6 Inches and a half long, a long Pocket-Fork; the Handle is Ivory, but is dyed of a very dark-brown Colour; the Iron Part is very black and smooth, but not rusty.

In 1865, fellow surgeon T. J. Ashton put together a whole list of foreign bodies that had been found lodged in people's rectums. There are enough here to start a well-stocked department store of household articles. 'Bottles, pots, cups, a knitting-sheath, a shuttle with its roll of yarn, a pig's tail, ferrules, rings, pieces of wood, ivory, metal, horn, cork, bone, &c.' You can't make this stuff up. There's nothing too preposterous that hasn't been found in an intestine.

> Nolet, surgeon to the Marine Hospital at Brest, relates the case of a monk, who, in order to cure himself of a violent colic, introduced into the rectum a bottle of Hungary wine, having previously made a hole through the cork to permit the fluid to flow into the intestine. In his desire to accomplish his object, he pushed the bottle so far that it completely entered the gut. Various means were tried to remove it without effecting the object; at last, a boy, between eight and nine years of age, succeeded in introducing his hand into the bowel, and withdrew the bottle.

The deadliest object turns out to have been a vegetable. In June 1842, a sixty-year-old man died on the way to the infirmary as a result of having eaten too many dried peas six days earlier.

> On examining the body after death upwards of a pint of grey peas was found in the rectum: they had been swallowed without mastication, and had undergone no alteration in passing through the alimentary canal, except becoming swollen by warmth and the absorption of moisture. The urethra was pressed upon, and he had had retention of urine for four days. The bladder was enormously

distended, its apex reaching the umbilicus, and its base nearly filling the brim of the pelvis.

Usually the obstacle at the back door is less foreign than you would wish it to be. Your body made it itself. It's your own fault. If your body were a cinema or a theatre, there isn't a fire brigade in the world that would tolerate such an obstacle. Fire isn't likely to break out in an intestine, not even after you've eaten the hottest peppers, but that doesn't make it any less painful. The name of the obstacle sounds ominous enough: haemorrhoid. A haemorrhoid is a varicose vein that's stuck in your arse. Haemorrhoids are the price we pay for the pride we take in walking erect. A respectable four-legged animal doesn't have varicose veins, let alone haemorrhoids, but if you stand up straight like a human being it means your heart is a long way from your toes, so the blood pressure has to be accelerated. This is more than the valves in the veins can bear. They tear, blood gets backed up, and your skin starts looking like a river delta. When this kind of twisted blood vessel forms in your anus it becomes fused with the intestinal covering and some of the connective tissue, resulting in a haemorrhoid. A quadruped like a dog is never bothered by such things. Dogs carelessly strew the streets with their shit until the day they die. A dog's backside is more or less the high point of its body. Its rear end is not only behind but also above its heart. Blood can never get stuck there. Our backsides are too low. There ought to be valves in the blood vessels in our bottoms, but in the process of learning to walk upright evolution forgot to include them. As compensation, humans have at their disposal a friendly member of the haemorrhoid family, the haemorrhoidal plexus, which, like its relatives, consists of twisted veins and can swell up with blood. It doesn't do this to torment you, however, but to seal off the last chinks in the closed anus.

You can dismiss a foreign object or a haemorrhoid as an incident that impedes traffic, like a jack-knifed lorry or temporary roadwork.

But the flow can also be blocked by the system itself, as in the case of bumper-to-bumper traffic, when it's the cars themselves that slow each other down. Slow-moving and stationary traffic in the intestines is called constipation. It's a self-reinforcing process. It begins with a sluggish turd. Because the turd stays in the intestines for a long time, a great deal of water is extracted from it. That makes it hard. Hard turds don't make any headway. They chafe against the intestinal walls and crunch at the curves. This causes new delays, more dehydration, congestion, and faces flushed from all the ineffective straining on the toilet. Shitting starts to look as terrifying as a breach birth.

Constipation is sometimes the fault of the intestine. The muscles are too slack, a narrowing has occurred, or the second brain has gone to sleep. That's a matter for the doctor. But just as it is with traffic, the fault lies more often with the people who use the roadway than with the roadway itself. Tough, misshapen turds hold up intestinal traffic. Eating fibre helps. When turds are softened by well-lubricated fibre, they move along at a nice clip and fill the toilet bowl, with time to spare.

Even if you haven't done anything wrong (as far as you know), sooner or later every drain gets clogged. And in one out of seven people constipation is a chronic condition. The symptoms are pain and straining, small pellets, low success frequency (fewer than three times a week), and the feeling after you've defecated that you're still not empty. Time for the drain cleaner. Laxatives can stimulate the movement of the intestinal wall or improve the lubrication along the way. Most of them work on the basis of osmosis. They might contain magnesium compounds that increase the number of dissolved particles in the fluid of the large intestine. Because the fluids on both sides of the intestinal wall strive for an equal concentration of particles, the laxative retains water in the intestine. That makes the turds nice and creamy. But it's not an ideal system. The extra water that is taken into the excrement causes valuable potassium salts to go to waste. A lack of potassium weakens all the muscles in the body, including those in the intestines, which ultimately results in the very constipation you're trying to treat. According to stomach specialist E. Mathus from the Academic Medical Centre in Amsterdam, what works best is a glass of lukewarm water first thing in the morning on an empty stomach, followed by coffee. 'With a cigarette,' he added in 1999, but even back then you couldn't say that. As an alternative to a laxative or cigarette, according to George Drewry a hundred odd years earlier in *Common-Sense Management of the Stomach*, there was always the enema.

> The regular use of the enema syringe is of the utmost importance in these cases; for, as I have already pointed out, nothing can be more irrational than the practice of taking purgatives by the mouth to remove hardened matter distending the lower bowel, when the enema syringe furnishes us with a means of removing the source of obstruction safely and comfortably in ten minutes, without the use of any medicine whatever.

Rabelais would not have agreed with him. He knew of an even older and better method for driving a firmly entrenched piece of shit from an obdurate body: fear.

> ...for one of the symptoms and accidents of fear is, that it often opens the wicket of the cupboard wherein second-hand meat is kept for a time. Let's illustrate this noble theme with some examples.
>
> Messer Pantolfe de la Cassina of Siena, riding post from Rome, came to Chambery, and alighting at honest Vinet's took one of the pitchforks in the stable; then turning to the innkeeper, said to him, '*Da Roma in qua io non son andato del corpo. Di gratia piglia in mano questa forcha, et fa mi paura.*' (I have not had a stool since I left Rome. I pray thee take this pitchfork and fright me.) Vinet took it, and made several offers as if he would in good earnest have hit the signor, but all in vain; so the Sienese said to him, '*Si tu non fai altramente, tu non fai nulla; pero sforzati di adoperarli piu guagliardamente.*' (If thou dost not go another way to work, thou hadst as good do nothing; therefore try to bestir thyself more briskly.) With

this, Vinet lent him such a swinging stoater with the pitchfork souse between the neck and the collar of his jerkin, that down fell signor on the ground arsy versy, with his spindle shanks wide straggling over his poll. Then mine host sputtering, with a full-mouthed laugh, said to his guest, 'By Beelzebub's bumgut, much good may it do you, Signore Italiano. Take notice this is datum Camberiaci, given at Chambery. 'Twas well the Sienese had untrussed his points and let down his drawers; for this physic worked with him as soon as he took it, and as copious was the evacuation as that of nine buffaloes and fourteen missificating arch-lubbers.' Which operation being over, the mannerly Sienese courteously gave mine host a whole bushel of thanks, saying to him, '*Io ti ringratio, bel messere; cosi facendo tu m' ai esparmiata la speza d'un servitiale.*' (I thank thee, good landlord; by this thou hast e'en saved me the expense of a clyster.)

Serious or not, this kind of advice has its origin in another era, and in more than one respect. From the mediaeval nobility to the French courtiers and the Victorian middle class, people were obsessed by constipation. That's hardly surprising when you consider what kind of food was being consumed: great chunks of meat, cauldrons of porridge, schools of fish, and broods of chickens, doused in heavy sauces and served in several courses. Intestines worked overtime until they finally threw in the towel. This did not leave their owners unmoved. While people of the twenty-first century are fanatical in their pursuit of health through healthy eating, those of earlier years hoped for a long and happy life by means of healthy evacuation. That's what their herbals were all about, that was the focus of their doctors' expensive advice. Gaunt adolescent girls were suffering from an intestinal problem, of course; anorexia nervosa was still an unknown condition. The focus wasn't on what went in but what came out. Of all the lists of medicines, the list of purgatives was the longest—with a preference for rectal over oral administration.

This anal phase in medicine reached a high point in around 1900. After the discovery of a large number of pathogenic microbes, all bacteria were regarded with suspicion. And the large intestine was full of them! If that intestine were to become clogged with shit, the bacteria there could use it to brew poisonous substances that would race through the blood and threaten the entire body. 'Auto-intoxication' is what Charles Bouchard called it in 1887—'self-poisoning'. Today we blame absence of exercise for everything, from baldness to cancer and dementia, but back then constipation was the mother of all maladies. It could lead to urinary tract infections, arthritis, headache, thyroid disorders, heart disease, feeble-mindedness, epilepsy and—oh, yes—cancer and baldness, of course. Eventually everyone would be able to tell by your skin that you were constipated, warned Dr Kellogg.

During the heyday of nudism, sandals, beating each other with birch branches and singing idiotic songs about the New Man ('Awake!'), a regular War on the Bowels was declared in the early twentieth century. Popular magazines, polite conversation and entire libraries were filled with advice on how to fight constipation. Enemas were pimped with rubber tubes and bellows, yoghurt and cornflakes performed their first abdominal miracles, and advertising agencies got rich from the laxative commercials ('If nature forgets, remember Ex-Lax'). True believers had their intestines shortened out of faecal fanaticism with the same ease that people today have their intestines re-routed round their stomachs as penance for their gluttony. But the intestinal trend faded with the Charleston after the roaring twenties. In 1930, a certain Dr W. H. Graves made a lot of noise with the claim that 90 per cent of all illnesses were caused by constipation, but it was too late. Looming on the horizon after the close of the anal age was the oral age. The rectum was promoted from 'a useless and superfluous structure' to 'part of a wonderful, multifunctional, and still incompletely understood digestive tract' with 'marvellously varied bacterial flora by means of which we become an integral part of the ecosystem'.

~

Shit can also be too thin. When that happens, that part of the magnificent, multifunctional alimentary canal has failed to understand itself completely and contains too much water. It's just as if you'd taken a laxative. And often that's exactly what happens. Besides the pharmacist, the greengrocer also sells outstanding laxatives, but without any instructions for use. Examples are apples, pears, prunes, rhubarb, blackberries, cherries, peaches and figs. And there's a very good reason why fruits like cherries tend to loosen your bowels. If a cherry is eaten by a bird, the bird quickly shits out the pit before it can cause intestinal damage, so it remains capable of germination. Charles Darwin cultivated all kinds of seeds taken from bird faeces in order to understand the distribution of plants. But even in human intestines many seeds retain their strength. Whenever there's a leakage of sewage containing human faeces, tomato plants sprout up without having been sown. Prune trees last longer, but the laxative effect is no less powerful. It really gets you moving.

If you can't get to a toilet on time, you may be in for a foretaste of what the future could hold for you: incontinence. Due to age or illness, a small percentage of the earth's population lose control of their pee, shit or intestinal gases. In the West, approximately 1 per cent of the population cannot hold their faeces. This is especially true of old women. Although something may be wrong with their muscles, it's usually the nerves that give out first. Many people wonder whether this will be their fate as well. It's all too easy to forget that we began our lives incontinent. A healthy baby shits and pees to its heart's content whenever its intestine or bladder feels like it.

So why do we bring these filthy babies into the world? Because it's been our habit for millions of years. The model for our babies comes from the time when our forefathers—the babies' fore-forefathers—were still living in trees as monkeys. Carefree as the birds of the air, they spent their lives at a splendid height and freely let their shit drop all over the poor earth-dwellers below. Those creatures are much tidier. They only shit in certain places, or hurry away when they're done. Now humans are earth-dwellers, too, and have wisely decided to clean up their act. But it isn't easy. Controlled defecation has to be learned, like playing the piano—and there weren't any pianos high up in the trees, either. Incidentally, many children master the piano before mastering their own sphincter. In the Netherlands alone, from 50,000 to 100,000 children attend the poo clinic because they keep having accidents. Or they hold their shit until a 'boulder' or 'elephant turd' forms and gets in the way of the excrement behind it, which somehow has to trickle through. It takes a lot of pills and even more careful attention before they're grown-up and toilet-trained. It may take years before a person gets the hang of this basic life skill. And even then, when things get very scary, or very funny, the occasional adult will feel something warm between their legs that quickly turns cold.

Full chamber pots have to be emptied. Clean ones look nice, but usually it's just a matter of moving the filth from one place to another.

Well shat and well forgotten. Every turd is heartlessly cast aside after its services have been rendered. Dumped. That's what a bottom is for. Any respectable animal has a head at the end where it's going and a bottom at the end where it's been. When it shits, it automatically leaves its filth behind where it won't trouble it anymore. It's a good system—until too many of your species line up in front of you and leave *their* filth behind.

That's called an environmental problem. For human beings it all began with the emergence of cities and villages. At one time we made sure we got away from our shit, but now we had to try to get our shit away from us. A little way outside the city or village was good enough for now. But what actually happens to a dumped turd? With luck, nature finishes it off in good time. But how? And what is good time? No one knew, because no one went to find out—until the 1960s, when the cow poo in Australia refused to decompose. Only then did it occur to anyone that even in the British motherland a cow pat would often lie there for six months at a time, much to the distress of the farmers. The fact is that a cow won't eat where it has shat, and the grass underneath becomes scorched from the excess of manure.

Apparently there's a big difference between tropical and temperate regions. In tropical regions, a lot of poo is carried away by dung beetles. During the rainy season they pop up out of the ground en masse to roll the dung into balls and carry it away. An elephant turd can disappear that way in a few hours. Dung beetles are attracted by the smell, as the Greek playwright Aristophanes already knew. In *Peace*, the hero Trygaeus goes to heaven on the back of a gigantic dung beetle in order to visit the gods. Fearing that the flight might be abruptly terminated, he asks his countrymen to make sure the beetle doesn't smell any manure as long as he's on board. He instructs his slaves to close off the privies and alleys with new masonry, and tells everyone to keep their arses shut.

Once a dung beetle has smelled its favourite odour, as Trygaeus knew, there's no holding it back. More than 2250 years later it had lost none of its gusto, as we see from *Souvenirs entomologiques* by Jean-Henri Fabre (1823–1915):

> What a lot of jostling for the same lump of manure. Never have fortune hunters, from every corner of the globe, applied themselves so diligently in the exploitation of a California gold mine. The pungent odour has spread the good news for miles around, and all have rushed in to claim part of the supply.

For a dung beetle, manure is its whole world, its everything, its present and its future. And its past: manure is its first memory as a larva in that underground paradise, surrounded by the fine shit so lovingly kneaded by its mother:

> The coarse loaf mixed with sharp stalks of hay may be adequate for her own needs, but she sets higher standards for her family's confectionery: it must be fine, exceptionally nutritious, and easily digestible. For this she needs sheep mortar, not the dark green olive-shaped kind that the wether scatters on the roads in rich abundance, but those that were fashioned to a soft cake by moister bowels.

Among the ancient Egyptians the dung beetle was sacred. They saw the beetle, pushing along a ball of manure, as a symbol of the sun in its orbit. But as Fabre already knew, there's nothing sacred about the scarab. As soon as his ball is finished, a neighbour who has just begun his own work 'abandons his labours to lend a helping hand to the fortunate owner'. It looks like an ideal collaborative effort, but it's not. 'It's nothing but attempted robbery, pure and simple.' 'The sympathetic colleague, under the pretext of offering a helping hand, nurses his heinous plan of taking possession of the ball at the first opportunity.' While the legal owner is busy digging his larder, the wicked companion quickly makes off with the ball, which he carries away with the speed of a pickpocket who's afraid of getting caught red-handed. If the ball is meant for its own consumption, then what comes next is a gala of eating—and of shitting:

> Twelve hours, or even more, is quite long for a session of gorging, but it makes for healthy and rapid digestion. While the material at the front end of the animal is continuously being ground up and swallowed, it re-emerges from the back, all in one piece, stripped of its nutritious components, and spun into a black thread, not unlike the waxed thread used by cobblers. The dung beetle defecates constantly during his meal, so fast is the operation of his alimentary

canal. For as long as his dinner lasts, the thin thread emerges unbroken and coils itself neatly like an easily untangled ball of yarn, as long as it hasn't begun to dry out. From time to time I pick up a piece of the thread with a pair of pincers and place it on a ruler divided into millimetres. The total of my measurements gives me the not inconsiderable figure of 2.88 m, produced in twelve hours.

In addition to the sacred variety there are 7000 other dung beetle species. Many are specialised in certain kinds of shit donation involving their favourite elements: one type prefers something juicy to suck on, the other goes for the harder bits, and a third loves to gnaw on dried crackers. In the desert, the dung beetles first soak rock-hard camel droppings in the damp earth, like biscuits in tea. Problems with manure processing once occurred in Australia, as mentioned earlier. There are no mammals there like the ones we have, only marsupials. Marsupials are fun to look at, and some of them are able to take enormous leaps, but this doesn't help the farmers very much. You can't milk them. So in 1788, humans brought cows to Australia. Besides milk, cows also produce cow pats. In the Netherlands these have been used as valuable fertiliser since time immemorial. But that didn't work for the colonists in this land of misfortune. Their grass wouldn't grow on cow manure. Left unused, the manure stayed where the cow had dropped it. So the little bit of grass that the rabbits had left alone was soon inundated. Given the fact that one cow could spoil 800 m^2 of pasturage annually with its excrement, Australia began losing grasslands the size of the entire Netherlands in a single year. In addition, there were all sorts of horseflies and other carriers of disease living in the cow pats. What ought to have been a land of milk cows and honey bees deteriorated into a land of shit and dung flies.

At the heart of the problem was the fact that grass doesn't live on manure but on the nutritive salts locked in it. The task of releasing these growth inducers falls to a great extent on the broad shoulders of dung beetles. But Australian dung beetles only like marsupial dung, which is nice and dry and fibrous. A floppy cow pat is not the sort of thing that makes their mouths water. This may testify to their good taste, but it left the Australians with a problem.

Just as the Dutch tackle their problems by setting up a new commission, so the people of Australia decided their problems had to be solved by importing a new species of animal. Two centuries after the arrival of the cow, dung beetles were brought in, this time from Africa. The beetles discharged their duties beyond all expectation. The cow manure was removed in no time. Yet that wasn't the end of the problems. The pathogenic dung flies remained. In their determination not to make any mistakes this time, the Australians had made the mistake of subjecting the beetle to a rigorous quarantine before being imported. So not only were the unwelcome germs left in Africa, but so were the mites that live on dung beetles. And it was these very mites, which are transported by dung beetles from cow pat to cow pat, that control the spread of dung flies best by eating their eggs. The Australians' next step, of course, was to import cow dung beetle mites. Don't ever get the idea that they're crazy down under.

There are more flies than beetles on a Dutch cow pat. Dung flies really love shit. Shit is their very meat and drink, their be-all and end-all. But not all shit is equally delicious. Just like the beetles, each species of fly has its favourite recipe. One goes for soft mush, the other thick gruel, and a third wouldn't mind the addition of something to gnaw on. Fans of pea soup or beef stew certainly must recognise this preference for a very specific consistency. Cow shit can often satisfy everyone's desire within just one pat, depending on the place and time. A cow pat gets juicier the further inside you go, and it gets drier the longer it lasts. That attracts a whole succession of species. The first to arrive are the horn flies (*Haematobia irritans*). They wait impatiently on the flanks of their benefactor until it gets the urge. Even before anything has made its appearance they assume their positions on the thighs, ready to plunge right into the virginal mush. According to expert G. A. Parker, studying mating flies on a cow pat may seem like a rather advanced form of perversion, but it is fascinating to see how within a minute the females have laid their eggs and are ready with dozens more, like football supporters in section C waiting for the next goal to be scored. Now the muck, no longer fresh, is attracting the other thirty-eight species of dung flies, one after another. At a distance they can tell from the smell if the correct degree of ripeness has been attained. Just like camembert. The males decide which pats are best for enticing the good females. If the cow pats get too dry, making it difficult for the females to penetrate the crust and lay their eggs, then only a few males will remain behind and the main force will continue on its way, to the next Promised Land. So new cow pats are always needed before the old ones are used up. But we can leave that to our livestock.

The fly larvae are no more able to polish off a Dutch cow pat than the dung beetle larvae or the earthworms, who are no slouches either. The greater part of a cow pat is broken down by bacteria and fungi, assisted by weather and wind, or trampled underfoot by cattle and pecked at by birds in search of larvae. In one gram of manure there are more than a million yeast cells and at least a hundred times as many bacteria. Compelled by the temperate climate to take their time eating, they spend an average of six months working on one cow pat. That's a life-saver for the parasites that exit the cow along with the shit. Because cows avoid places where they've already shat, it can take a long time before their parasites happen upon a new host. To increase their chances, the larvae of the nematode *Dictyocaulus viviparus* climb onto the sporangia of the *Pilobolus crystallinus* fungus. When the spores inside are ripe, the larvae take a free ride with the sporangia, which are shot into the air. *Dicrocoelium*, a sheep parasite, works its way into the body of a passing ant. This causes the ant to become confused, and it climbs onto a blade of grass. When it bites into the grass, its jaws clamp down, whether the ant wants them to or not, and it just hangs there, giving the parasite plenty of time to climb on board a passing sheep.

In the nineteenth century it became clearer than ever that cities could no longer leave their rubbish removal to hanging ants, sacred beetles and randy flies. London, Paris, Berlin and Amsterdam stank to high heaven. One cholera epidemic after another broke out, leaving many thousands dead each time. According to the then commonly held miasma theory, illnesses were caused by vapours, mist and evil-smelling gases. Koch and Pasteur would set us straight. But even before their time, in 1849, it was known that cholera had something to do with water polluted by excrement. Dr John Snow had carefully mapped this out for London. Vapours, dirty water or germs: when it came to fighting the disease, it didn't much matter how it was explained. The shit had to go. And more hygiene had to be introduced. Hygiene was not a new virtue; it was an old emergency measure necessitated by overpopulation.

How do you get rid of shit? Many people had cesspools. When mixed with household rubbish and ashes from the hearth, human waste proved useful as fertiliser. For centuries you could even earn good money with it. But emptying the cesspools was not a pleasant job. Often the brickwork was too loosely constructed, and deliberately. The thinnest stuff would seep out and run into the soil, so you wouldn't have to empty the cesspool so often. Anyway, it's hard to dig a cesspool when you're three storeys up. A more convenient solution in the inner cities was the bucket system. You shat in a bucket, and when that was full it was picked up and replaced by an empty one. In Frisian cities such as Leeuwarden and IJlst this system was still in use in 1970. Apparently it was still cost-effective in such agrarian districts. But that didn't make the stench or the nuisance any more tolerable.

Overpopulated cities were an outgrowth of the Industrial Revolution, which gave rise to solutions as well as problems. A century before Leeuwarden got rid of its poo buckets, the Dutchman Charles Liernur came up with a brilliant idea. Using the most modern steam engines and

Changing the shit barrels in Haarlem.

pneumatic pumps, he could pump the faeces and urine out of a private toilet and into a system of cast iron pipes. The muck would be taken to underground collections points, from which it would be driven out of the city. Hendrik, prince of the Netherlands and governor of the Grand Duchy of Luxembourg, was immediately enthusiastic. It was partly thanks to him that the Liernur System was adopted in Amsterdam and Leiden in 1870. A program of furious digging began to lay the pipes and excavate the pits. All the citizenry had to do was shit.

> Then a traction engine with a steam-powered pneumatic pump drives to the small underground street reservoir to pump out the air from both the reservoir and the whole system of main and subsidiary pipes, up to the hermetically sealed house valves. These are then opened and shut one by one so the contents of the toilets, gasses and all, can be discharged into the street reservoir. Once the vacuum is complete, or almost complete, a column of air sweeps in when each house valve is opened that has a mechanical force roughly equal to that of thirty hurricanes.

You didn't have to flush. The special Liernur Toilet Bowl was fashioned at such a steep angle that the shit never even came in contact with the back. Any filth that landed on the front would be automatically rinsed away during urination. Urine also kept the pipes clean free of charge; the 'greasy components in the faeces' would protect them from rusting. In actual practice, however, the system was rather disappointing. Cast iron, rubber and steam were unable to cope with faeces and urine. Not only did the system require a lot of maintenance, but pumping out the cesspools every day and carrying off the waste was also very labour-intensive. In 1912 Amsterdam was the last of three cities to give it up.

Like the Arsehole of Rome, the Cloaca Maxima emptied into the Tiber.

There was nothing wrong with Liernur's concept. He was just too far ahead of his time; the technology wasn't ripe. Today's sewers also use a great deal of air pressure, not to pump out but to press. Back in the day, Liernur and his air had to give way to water. In spite of the Industrial Revolution, people all over the world opted for a concept from antiquity: the flush system. The great pride of this system was the Cloaca Maxima, the Great Sewer of Rome, to which even public toilets were connected. Starting in the sixth century BC, the Romans simply covered over the rivers and streams that were already being used to carry away waste, and they expanded the system with underground brick canals. At about the beginning of the current era, the Greek traveller Strabo described the Roman sewer as 'large enough here and there to allow a wagon full of hay to pass through it; while the inflow of water via the aqueducts was so abundant that it looked as if entire rivers were streaming through the city and the sewers'. The sewer builders must have looked to the Greek myth of Heracles for inspiration. Heracles had been ordered to muck out the stables of King Augias. He was given one day to clean up thirty years' worth of shit from three thousand cows. To do this, he made two holes in the stable walls and let two rivers run in, the Alpheus and the Peneus. But when Heracles went to collect his wages the king refused to pay him because the actual work had not been done by Heracles but by the rivers.

The Romans decided to let the Tiber do the dirty work. Two thousand years later, London handed the job over to the Thames, Paris to the Seine, and Amsterdam to the Amstel and the IJ. Each time the same solution was used to tackle the same problem. But geography wasn't always cooperative by any means. London is too far from the sea for drainage to be effective, and the waste was regularly washed back in on the following tide. Paris had more luck with its Seine. And when the crowded centre of Paris was being opened up to accommodate the construction of the sewers, George Haussmann took advantage of the situation to give the city a new face. He himself was delighted by Paris's new backside:

> Like organs of the great city, the sewers will work just like those of the human body, without ever seeing the light of day. The pure and fresh water, the light and the warmth will flow through them like the various liquids that benefit life with their active support. The fluids will do their mysterious work without ever disturbing the good operation of the city or harming its outer beauty.

In Holland the geography was downright hostile. In the absence of differences in elevation, the drainage water stagnated in the ditches and canals. The Dutch canals were dug partly as defensive structures and usually in concentric circles, as in the case of Amsterdam, which did not contribute to effective drainage. Amsterdam's main canals could be drained a bit with the help of the tides on the Zuiderzee and an ingenious system of sluices, but in the Jordaan district during the nineteenth century the stench was unbearable. There all the privies emptied into the canals, which were more or less stagnant. Tourists had to hold their nose. 'If there's one thing the Dutch really need to clean,' wrote the brothers Edmond and Jules de Goncourt after a visit in 1861, 'it's their water.' Amsterdammers who could afford it traditionally fled to country houses along the Vecht or the Amstel Rivers during the summer, when the stench was at its most extreme. When they returned in the autumn from Diemen or Ouderkerk south of Amsterdam the stench was there to welcome them. But the Department of Public Works, which had been set up in 1850, was already working on a solution. Canals were filled in and replaced by sewers. Yet it would be more than a century before all drainage into the remaining canals was halted. In my youth in Amsterdam, while skating on the Herengracht or the Keizersgracht in the winter, I would see warm steaming turds floating in their own little holes in the ice along the quay where the rich people had done their dumping. Today those turds go through sewer pipes—most of which run along the same canal but under water, tidily hidden from view. This is not to say that you don't see turds anymore in the city. Amsterdammers have dogs, too.

7

Brown Gold

Its odour is redolent of fine tobacco, the woodwork in an old church, sandalwood oil, the Wadden Sea at ebb, straw bedding in a zoo or a blend of freshly mown grass, the damp forest smell of a dead fern and violet perfume. Kings and emperors gave it to each other as a sign of supreme exclusivity. It was an essential ingredient of elixirs. Casanova drank it in his chocolate mousse as an aphrodisiac. When it comes to mythical standing, it can hold its own with black truffles and unicorn horn. Scholars racked their brains trying to figure out what it was. It was shit.

The shit of all shits, it ranks right up there with nectar and ambrosia. It is ambergris—grey amber. Not yellow amber, the hardened resin, but the grey amber that is found along distant coastlines (if you're extraordinarily lucky) and is said to have been vomited by a whale. But it isn't vomit. It's poo.

For more than a thousand years, ambergris has been sought after by the poor as a way of getting rich and by the rich as a way of putting on a good show. It's considered a fragrance from heaven. Yet in its pure form it has no smell at all. Ambergris seizes on other fragrances and gives them a deep lustre. It is still sought by the perfume industry, which pays 15,000 euros a kilo for it. Synthetic amber doesn't come within miles of the natural product. Shit.

The secret of ambergris was unravelled during the heyday of whaling. Herman Melville was already well aware of it in 1851 when he wrote *Moby-Dick*. In that tale, Captain Ahab and his crew come across a French ship, the *Bouton de Rose*, with two rotting whales in tow. The animals had died a natural death and now stink to high heaven. They are, in whalers' terms, 'blasted'. Second mate Stubb informs the French captain 'that only yesterday his ship spoke a vessel, whose captain and chief-mate, with six sailors, had all died of a fever caught from a blasted whale they had brought alongside', and implores him, if he values his life, to cut the towing cables. The Frenchman falls for it. He's barely gone before Stubb climbs on top of the smaller of the two whales.

> Seizing his sharp boat-spade, he commenced an excavation in the body, a little behind the side fin. You would almost have thought he was digging a cellar there in the sea; and when at length his spade struck against the gaunt ribs, it was like turning up old Roman tiles and pottery buried in fat English loam. His boat's crew were all in high excitement, eagerly helping their chief, and looking as anxious as gold-hunters.
>
> And all the time numberless fowls were diving, and ducking, and screaming, and yelling, and fighting around them. Stubb was beginning to look disappointed, especially as the horrible nosegay increased, when suddenly from out the very heart of this plague, there stole a faint stream of perfume, which flowed through the tide of bad smells without being absorbed by it, as one river will flow into and then along with another, without at all blending with it for a time.

'I have it, I have it,' cried Stubb, with delight, striking something in the subterranean regions, 'a purse! a purse!' Dropping his spade, he thrust both hands in, and drew out handfuls of something that looked like ripe Windsor soap, or rich mottled old cheese; very unctuous and savoury withal. You might easily dent it with your thumb; it is of a hue between yellow and ash colour. And this, good friends, is ambergris, worth a gold guinea an ounce to any druggist.

Stubb knew where to look for it, and it wasn't in the head. In every sperm whale there's a miraculous substance called spermaceti: a clear oil that, when exposed to the air, congeals into a soft white wax that is sold as salve in the pharmacy. For centuries, spermaceti was thought to be the seed of the sperm whale, hence the animal's name.

Spermaceti could fetch a great deal of money, but not nearly as much as ambergris. To find that you'd have to look in the intestines. Essentially, it's a big lump of encrusted old shit. The little specks you see in it are the beaks of the squids that the sperm whale has eaten. With its four stomachs, a sperm whale makes quick work of a squid's soft tissue, but it can't reduce the hard mouth parts, the eye lenses, and the internal shell. A healthy sperm whale vomits the beaks by the thousands once a week, like an owl vomits its pellets, but if there's anything wrong with the animal's valves and sphincters the beaks may pass on from the first stomach to the intestines, where they irritate the wall. As an oyster covers an irritating grain of sand with mother-of-pearl, so the sperm whale encapsulates the beaks in a fatty, cholesterol-rich secretion. The next load of shit gets stuck behind it and also begins to clot. Constipation like that must cause the sperm whale great inconvenience. If it's lucky it will shit the wad loose in plenty of time, but most of the animals get more and more constipated until their intestine bursts and they die. The body itself feeds the sharks and worms of the ocean, but the lump of ambergris floats to the surface. Its career has only just begun. Fresh from its master, the ambergris is still a pooy lump of tar. In the ocean it doesn't reach its full maturity for years, sailing on the surface, blown forward by the wind, carried back by the tides, exposed to air, light, salt and waves. Growing smaller and smaller, lighter and smoother, it finally washes up on a distant shore and lies there like a pale stone that's been through the laundry, waiting for the beachcomber who's lucky enough to make the discovery of his life and smart enough to recognise it. His nose will be rewarded with a fragrance that makes him think of all sorts of things but is impossible to describe. 'There is simply nothing else that smells quite like it,' writes Christopher Kemp in *Floating Gold*. 'It is like a single, remote point on a map with no landmarks anywhere by which to find it.' Actually, ambergris smells like just one thing: ambergris. Only biologists believe they can detect the faint faecal odour; all everyone else smells is money. In 1693, the Dutch East India Company paid the king of Tidore 11,000 dollars for a lump weighing 83 kilos. The lump was exhibited in Amsterdam as a rarity for many years and was finally broken into pieces and sold at auction, 'so that many persons now alive have been witnesses to it'. A find of exactly the same weight occurred in 2012 in a sperm whale stranded on the Razende Bol, a sandbar off the coast of the Dutch island of Texel. Although ambergris obtained from a sperm whale is less valuable than a loose lump washed up on the beach, the finders from the Ecomare seal sanctuary counted on a take of hundreds of thousands of euros.

Whether shit smells like money or not depends on your nose. While a city nose only lusts after ambergris in the carefully measured form of faecal fragrances known as perfume, a country nose realises that shit doesn't have to come from a sick sperm whale to be worth money. That nose will take healthy pig muck and cow shit any day. Although crops can live from air and light, they need substances from the soil, especially nitrogen and phosphorus, to build up their cells and their chemical machinery. These are just the substances that cattle would sooner be rid of than rich in.

When you forget to enrich the earth with excrement it soon becomes less productive. So the emergence of the mixed farm was almost inevitable: animal manure was used to feed the plants, which the animals then ate in order to make new manure. Manure is no saint, says the farmer, but it performs miracles wherever it lands. And the most divine plants spring forth from the filthiest shit. Barren deserts become fertile soil, depleted fields bear new fruit. In terms of miracles, cow shit and pig muck are in no way inferior to the finest ambergris. The shit that cows and pigs produce is worth more than the capital of all the farm credit banks put together. The value of any farm has always been found in its manure pile and not in its money box.

I learned that as a child. Back then I had a little donkey. Not a real one, but a toy made out of tin—you know, with those lips. When you pulled its tail, it shat out your savings coin by coin. The donkey was based on the fairy tale from the brothers Grimm in which a donkey shits gold coins at the word 'bricklebrit'. It's not for nothing that this image remains engraved in the collective memory. For a donkey, money is shit; for a farmer every turd was golden.

For centuries, shit determined the prosperity of the farmer. There was never enough of it. The more shit, the greater the yield. A farmer with more land was richer, mainly because a larger piece of land produced more manure. Air, light and water are generally available in abundance, but nitrogen and phosphorus have always been the restrictive factors, so they were worth money. Those who thoroughly understood that were the English liege lords. Their tenants were usually obliged to pen their flocks of sheep on the lord's land, so the most important revenues automatically went to him. He who had the power, had the manure; he who had the manure, had the power.

Because everyone used it on their own land, manure was hard to come by in many places, not even for money—until enormous amounts of the stuff were discovered abroad. Bird manure from South America. Guano.

Birds fly. Freed from their ties to the earth, they acknowledge gravity only when it suits them: when they have to shit. Usually they just let everything fall—on creatures like us, shackled to the earth's surface by the same force that liberates the birds from their burdens. This is hard on our coats and our cars. But the amount of damage pales in comparison to the value of the corrosive excrement. Bird manure contains thirty times more nutritive salts than normal stable manure. Yet collecting all the shit from tits and coots is a pointless task. Better to take note of where the birds gather. The heron colonies of Amsterdam are a good example of how much a bunch of birds can shit when they get together. But even these Amsterdam shit-herons have to acknowledge the superiority of the seagulls, gannets, terns, and pelicans of the tropical coasts.

Europe was given a taste of their potential in 1804 when the explorer Alexander von Humboldt came back with a sample of shit from the Chincha Islands of Peru. The rocks there were covered with a layer of guano that was 200 metres thick in places. Apparently the birds had plenty to eat in the surrounding waters, which were teeming with fish. And whatever excrement landed next to the island was more than enough to keep up the fish population. The shit fed the plankton, the plankton fed the fish, and the birds were bursting with minerals from the fish they ate. The minerals came primarily from the rocks on which the birds nested, which were slowly being washed away by the sea. But the shit on the rocks accumulated because there was too little rain to rinse it off. In the sun the shit became mummified. It mixed with seaweed, fish remains, pebbles, shells, and dead birds to produce the end product: guano.

Beyond South America there were also islands in the Pacific and along the coasts of South Africa where reefs rich in birds were being exploited. In 1856 the United States greedily proclaimed the Guano Islands Act:

> Whenever any citizen of the United States discovers a deposit of guano on any island, rock, or key, not within the lawful jurisdiction of any other Government, and not occupied by the citizens of any other Government, and takes peaceable possession thereof, and occupies the same, such island, rock, or key may, at the discretion of the President, be considered as appertaining to the United States.

By 1840, after a hesitant beginning, guano had also become a booming business in Europe. Two years later, 283,300 tonnes were shipped to Liverpool. Farmers were enthusiastic. G. Rijnders, a gentleman farmer at Groot Zeedijk in Groningen, concluded by way of experiment that one tonne of guano was as effective a fertiliser as 30 tonnes of horse, cow or pig manure and much cheaper to import than night soil from the big city. It was considerably more difficult to find people who were willing to harvest the guano in those remote regions. On Midway, one of the islands that the US had confiscated under the Guano Islands Act, the few inhabitants complained of having the feeling that they were living in a birdcage, 'except there's no one here to change the sand'. Elsewhere even the sailors refused to go ashore. Their eyes had started bleeding from the ammonia vapours. To meet the demand for guano, natives were recruited and coolies brought in. In *Caesar's Passage*, which takes place in around 1864, Andrew Smyth paints a picture that might have been taken from Dante's *Inferno*:

> The Chinese labourers work on the sheer cliff faces of white guano, digging and bagging amid the dust and overpowering smell of the deposits. Each gang was worked by an overseer with a twenty-foot whip which was used skilfully and instinctively, without thought or mercy. Many of the coolies threw themselves into the sea rather than continue with the hopeless misery of their situation.

The best guano doesn't come from birds but from bats. Bats live by the hundreds of thousands in caves, the floors of which are eventually covered with a layer of shit several metres thick. A century ago in America, special bat towers were built to make the harvesting easier. Dr Campbell's Bat Roost, built in 1918 in San Antonio, Texas, yielded 2000 kilograms of guano shat by 250,000 to 500,000 guano bats (*Tadarida brasiliensis*). The good Dr Campbell's main objective was combatting the malaria mosquito. And indeed, the incidence of malaria around the tower did decline. But when the manure at the site was examined in 1926 it was found that the bats had not eaten any mosquitoes. Malaria had disappeared even without the presence of bats. The towers were not worth maintaining for their manure alone, and they fell into disrepair. The finest memorial to the keeping of bats is, curiously enough, in Limburg in the Netherlands. Still standing on the De Bedelaar country estate is the bat tower of Eugène Dubois, once world famous as the discoverer of Java Man. The tower is probably the only shithouse in the entire country that has been declared a national monument.

Even though the guano in the New World was there for the shovelling, it had yet to reach the harbours of the Old. Coolies perished miserably while loading the ships, which were wrecked as they rounded Cape Horn. Was there no extra manure to be found in Europe itself? Where in Europe could you find such concentrations of shit-producing animals as on the rocks of the sea birds and the bat caves of America? Of course: in the cities. That's where human beings were packed together cheek by jowl. No colony of birds or bats could match the shit-producing capacity of the human masses of London and Paris in the nineteenth century. In 1871 a Native Guano Company was founded to capitalise on the anuses of the British. In the *Leisure Hour* of 1885, the initiative was supported by a fierce appeal:

> European nations send fleets to the New World to bring over costly guano, while neglecting wealth at home. It is not pleasant to think of the sewerage of a great city: but to the eye of science, when opened, this uncleanness means green grain and flowery meadows; it means thyme and marjoram and all fragrant herbs; it means

sweet-smelling hay and golden harvests; it means bread on your table and meat in your larder; it means warm blood in your veins; it means health, and joy, and life! The fertilising of the fields becomes the nourishment of men.

With approval Victor Hugo writes in *Les Misérables*:

Paris throws into the water every year twenty-five millions of francs! This is no metaphorical statement, but simple truth. How, and in what fashion? By night and by day. With what purpose? With no purpose. With what thought? With no thought. To do what? To do nothing. By what agency? By the sewerage of the city.

Sewers are insane. That became clear to me when I was living in student digs without a proper toilet. You could pee in the sink, but for shitting you had to leave the room, which my girlfriend was unwilling to do. She shat on an old newspaper, which she neatly folded up afterwards and tossed into the rubbish bin. That was something I never dared, and now it's no longer necessary. But of course my girlfriend was right. Tossing your waste in the bin is more logical than sending it into the environment with a tank full of valuable clean water through kilometres of pipe, polluting everything it comes in contact with along the way. The fact that the sewers today no longer flow into waterways or seas is great for the waterway or sea, but it does nothing to reduce the insanity or the insane amount of work involved. When the shit gets to the purification plant between the toilet and the waterway or sea it undergoes all sorts of technical contortions to separate it once again from the water. The water can be used to flush away new shit; the old shit is incinerated along with the household refuse.

Rubbish should be separated at the source. I obediently set aside my old paper, metal, garden waste and leftover paint in order to throw them into the appropriate bins at the recycling station in the city of Weesp. But there's no bin for my shit. Nobody wants it. Whatever became of the days when the dunny-can man came to pick up your excrement? What kind of madness is it, to withhold from the economy a source of energy that is many times more nutrient-rich than cow manure? This may end up being very costly to the West. In the East they're not so crazy. All through the centuries, both the Chinese and the Japanese empires were built on men's and women's shit.

Until the rapid urbanisation of the twenty-first century, more than 90 per cent of all the human faeces in China ended up in the fields. Every turd counted as one more. But you couldn't just squeeze out another one whenever you needed it. Your family did everything they could, too. Even travellers' excrement was looked on longingly. In *Aux Etats de Jersey* (1853), Pierre Leroux tells the story of how he and eleven other French travellers in Macau saw a 'smiling, vigorous Chinaman' approach their innkeeper to engage in negotiations. A translator explained to Leroux what was going on:

> What he is doing here is that he comes to buy the *niao* and *fenn* of your lordships. The *niao* and the *fenn* are more valuable than yellow gold. They feed the yellow crops that make man grow. Yellow gold only makes the hair grow white.

The innkeeper didn't get as much for the excrement from the twelve Frenchmen as he did for that of the six previous guests. But they were Englishmen.

> The great *Tien* gave the English singular faculties. One Englishman can fertilise a field that feeds a whole family. An Englishman is an animal that is always eating and is eating mainly meat. Happy are those on whose property they deposit their bounty. My father, who was a noted gardener, asserted that it took three pigs to match an Englishman, and that eight Portuguese were barely comparable...

The prestige of human manure reached a high point during the Cultural Revolution. According to Mao, your turd did not belong to you but to your entire commune. Your comrades dragged your excrement around, and you dragged theirs. As Ralph Lewin reports, 'As I can attest from personal experience, a pair of sewage-filled buckets, carried on a yoke over the shoulders, is no mean burden.' Fortunately he didn't have to lug it very far. 'My Chinese hosts were afraid I might spill some of the valuable contents.'

The extent to which the use of one's own manure is culturally linked was shown by anthropologist Sjaak van der Geest during the sixties, when the Chinese built the railway between Tanzania and Zambia. In accordance with custom, they fertilised their little gardens with their own excrement, much to the horror of the local Africans. 'The good yields from the Chinese gardens were not enough to change their minds.' This aversion to human shit has not worked in Africa's favour. At one time there was little need for manure in Africa because every time the soil became depleted you could burn the next bit of forest to the ground. But all this has changed with overpopulation. Fortifying the emaciated continent with cow manure hasn't worked due to a lack of cattle, which in turn is connected to the depleted grazing land.

What do all those tribes from the West, the North, and the South have against their own human manure? Is it that it's dirty? Cow manure is dirty, too, yet we spread it on our crops with a whistle on our lips. The taboo has less to do with shit than with its human origins. The shit from a member of your own kind is taboo, just as the meat of your own kind is taboo. The aversion to shit and the aversion to cannibalism are based on the same thing: fear. Fear of disease. And that is not groundless: you have the most to fear from your own kind because they house the same evil. This was proven once again in 1959, when a mysterious trembling sickness was discovered among the Fore people of Papua New Guinea. The victims had all eaten the brains of their deceased fellow tribesmen during tribal rituals. In doing so they ingested a prion (infectious protein) that attacked the central nervous system and in most cases caused death within a year. This disease of human flesh is nothing compared with the epidemics unleashed by human faeces. For cholera bacilli and other pathogens, polluted drinking water is the fastest way to travel from backside to mouth and to backside again.

Human excrement is scary, but that's no reason to throw it away. It's better to tame it. Turn it into real manure. Manure is tamed shit. To turn a wolf into a dog you have to domesticate it, and to turn shit into manure you have to compost it. This involves letting the creatures you fear do all the work: bacteria, fungi, worms and other tiny beasties. They turn scary excrement to beloved manure. And while they're at it, they clear away the malicious relatives. The only thing you have to do to get them working is to mix the shit with other organic rubbish and put it in a heap. In no time at all, the temperature in the heap will rise to levels that no normal pathogen can tolerate. But certain microbes can tolerate it. As true thermophiles they actually thrive at temperatures of sixty or seventy degrees Celsius, which they themselves have generated by their activities. After a few weeks there may be some worm eggs that are still alive, at the most. The thermophile bacteria are past their prime by then and ordinary bacteria take over, helped by mites, fungi and insects. If you let the heap sit for one or two years, you end up with the most scrumptious food a plant could ever want.

'The Lord giveth,' the church teaches, 'and we make of it what we will.'

To keep from having to take every turd to the compost heap as soon as it's deposited, you collect your shit at the bottom of your compost toilet. These come in all shapes and sizes. Not all of them reach the temperature required to kill pathogens, however. One that does is the Humanure toilet developed by pioneer Joseph Jenkins. Instead of being flushed away, every portion of excrement is covered with a layer of peat or sawdust. When the reservoir is full, the contents are moved to the actual compost heap, located elsewhere. After one or two years you can enjoy the satisfaction of eating tomatoes that are self-pooed, as it were. Not all of Jenkins's guests have been able to appreciate this, however:

> A young English couple was visiting me one summer after I had been compositing humanure for about six years. One evening, as dinner was being prepared, the couple suddenly understood the horrible reality of their situation: the food they were about to eat was *recycled human shit*. When this fact abruptly dawned upon them, it seemed to set off an instinctive alarm, possible inherited directly from Queen Victoria. '*We don't want to eat shit!*' they informed me, rather distressed (that's an exact quote), as if in preparing dinner I had simply set a steaming turd in front of them with a knife, fork and napkin.

The English couple are not alone. Most people don't see the value in their own excrement. Yet this is nothing compared with the human muck on which the mediaeval cities and the Chinese Empire flourished. But if you were to fill up a bucket with your shit today you'd have a hard time getting anyone to buy it. If shit is the currency by which the value of agriculture is expressed, then its exchange rate has plummeted drastically. It all began with the German chemist Justus von Liebig (1803–1873), who discovered that crops grow quite nicely without organic fertiliser. Mother Earth does not possess any magical life force; a plant will thrive just as well in a vase with the proper saline solution as it does on well-fertilised humus. So farmers were no longer at the mercy of human or animal faecal capacity. Dunghills made way for sacks of artificial fertiliser. In 1850 the first crops were fertilised with superphosphate, a salt you could mine like coal. Phosphate is good for the growth of plants. It supplies them with building materials and stimulates the absorption of solar energy. You need potassium for the synthesis of carbohydrates, which constitute the bulk of beets and potatoes. Potassium became cheaply available in around 1860 as a waste product of the salt mining process. But the most urgent need was for nitrogen. The ammonium sulphate that was released as a by-product in the production of coke after 1890 couldn't keep up with the demand, even though nitrogen is almost omnipresent. Four-fifths of the air we breathe consists of nitrogen. The problem is that our crops cannot extract it from the air as a gas (N_2). A plant doesn't inhale nitrogen, it drinks it, with its roots. But that means the nitrogen has to be available in soluble form. The only life forms that can convert insoluble nitrogen gas into soluble nitrogen salts are certain kinds of bacteria. Fortunately the soil is crawling with them. The bacteria themselves can't be seen with the naked eye, but you can see the little tubers on the roots of peas and beans in which they live. If you plough these plants under you'll be extracting the nitrogen from the air, which you can use as food for other crops. But a natural process like this can't keep up with the hunger of modern agriculture. Now it was all a matter of waiting for a trick to come along by which the work of the bacteria could be done in a chemical factory.

It didn't take long. The First World War broke out. The need of food for friends was surpassed by the need of explosives against enemies. The best explosive, TNT (trinitrotoluene), consists chiefly of nitrogen salts. Back then the main source of these salts was still guano. The First World War was in danger of being fought with bird shit. Scientists from either side searched for other sources, but the battle was finally won by the Germans, thanks to the chemist Fritz Haber (1868–1934). In 1908 he succeeded in combining nitrogen (N) with hydrogen (H) to produce ammonia (NH_3). Not long afterwards, Carl Bosch (1874–1940) invented a reactor chamber in which you could do this on a grand scale and make as much ammonia as you wanted, the raw material for artificial fertiliser and bombs. A global famine due to a shortage of nitrogen was averted, and a world war could be staged in all its intensity.

After the First World War, Haber was no longer welcome in France, which had suffered terribly from his TNT. Nor was he welcome in Germany, since he was Jewish. Physicist Max Planck tried to put in a good word for him with Hitler, but to no avail. 'If the dismissal of Jewish scientists means the liquidation of modern science,' Hitler answered, 'we'll just have to go a couple of years without science.' After fleeing to Switzerland, Haber saw his discovery result in millions of victims for a second time. Yet he himself lives on in the public memory as 'the man who made bread from the air'. The Haber-Bosch process is still the most important source of nitrogen for our crops, even more important than all the land bacteria in the world. One-third of the world's population is fed thanks to Haber's ideas.

You've got to keep the customers satisfied. No one has to go hungry anymore for want of nitrogen, or to mess about with cow manure and human muck. But living happily ever after continues to evade us. Artificial fertiliser is a Trojan horse. While manure used to be a source of income, today's fertiliser has to be paid for. And making bread from the air requires an enormous amount of energy: the Haber-Bosch process only works above 500°C and at a pressure of hundreds of atmospheres. Not only that, but once it's in the soil artificial manure does bad things as well as good. A lot of it leaches into the ground water, where it contaminates the environment and threatens the water supply. Nor does artificial manure do everything that natural manure does, not by a long shot. A sack of salt doesn't have the structure of farmyard or human manure, which releases its nutrients gradually and provides a good home for insects, fungi, worms and other life forms that improve the structure of the soil. The little turds left by earwigs and centipedes also produce manure that's just as good as the cow pats covering their heads, and their farts provide aeration. A beaker of saline solution may be just as effective in the laboratory, but in the garden only organic fertiliser can preserve the soil from depletion and crumbling over the long term. The flowers in the garden of Mr and Mrs Pratt even asked for it, according to cabaret artist Jasperina de Jong:

> Mr Pratt and Mrs Pratt,
> We want fertiliser, no problem with that,
> But potassium sulphate? It makes us recoil
> Since slowly but surely it ruins the soil,
> Listen, we beg you, and open your eyes,
> Only natural manure is healthy and wise,
> If not from the cow, then at least from the cat,
> Mr Pratt and Mrs Pratt.
>
> What says the perennial species?
> 'I want faeces.'
> What says the rose in the garden of the señorita?
> 'I want excreta.'
> What says the lily so chaste?
> 'I want solid waste.'
> And what about the crocus so young?
> 'Oh, how I yearn for good honest dung,'
> Glorious dung, glorious dung.
>
> Mr Pratt and Mrs Pratt
> We demand fertiliser, no problem with that,
> But potassium sulphate? Are you still unaware
> That it harms our well-being and poisons the air.
> It's downright insulting, a blight to our eyes,
> Only natural manure is healthy and wise,
> If not from the cow then from where you last sat,
> Mr Pratt and Mrs Pratt.

You can tell what artificial fertiliser does to the soil by looking at the worms. In a healthy pasture, there are as many kilos of worms living under the ground as there are kilos of cows living above it. They're trifling little creatures at first glance, good enough to serve as food for blackbirds or bait for fishing. But not too shabby for Charles Darwin. He devoted his life to them. It wasn't his theory of evolution that occupied him the longest, but his study of worms: from 1837, when he gave his first lecture on them, until shortly before his death in 1881, when his last and most charming book came out, *The Formation of Vegetable Mould through the Action of Worms*. 'It may be doubted,' he wrote, 'whether there are many other animals which have played so important a part in the history of the world.' Earthworms have not inherited the earth, as we humans have; they have created it. With shit. Like a living tube, a worm regularly squeezes out the contents of its intestines from below the surface of the field in the form of the well-known little piles of pasta. Although such a pile weighs only ten grams, all the earthworms taken together excrete five kilos per square metre a year. And it's excellent manure. It contains much more nitrogen, phosphorus, potassium and calcium than five kilos of ordinary soil. By shitting, the earthworms make the minerals from the soil available to the plants. Aristotle called them 'the intestines of the earth', and for very good reason. If you're a farmer or gardener and you want to harvest healthy crops, you'd be well advised not to fertilise the plants but to feed the worms.

With his study of worms, Darwin validated his idea that small changes can have great consequences over the long term. Bite by bite, shit by shit, all the soil of a forest or field makes its way through the intestines of the worms once every three years. The forest path you walk on, the field on which your cauliflower is grown, all the stuff we call earth, is brown for a reason. It's shit, shit from earthworms and from all those other animals who have to work outside without the benefit of facilities. They cover everything with it. At one point Darwin scattered fragments of lime about as evenly as he could, and twenty-nine years later he found them at a depth of dozens of centimetres. Entire houses can disappear over the course of time. Even the colossal stones of Stonehenge were buried by worms. Archaeologists have their hands full trying to recover Roman villas that have been completely obliterated by faeces. At such an excavation in Surrey, Darwin saw worms burying stones that had just been uncovered. If you look at it this way, the entire country is one big dung hill. And in the sea it's even worse. There's not a toilet in sight, not even a tree to pee against. Fish, shrimp, sea cucumbers, jellyfish and mussels—not to mention whales—release all their filth right into the water, and this goes on day in and day out, year in and year out, eon after eon.

The earth that you cherish turns out to be worm poo, the sea where you spend your holidays is filled with fish piss. What we call the environment is nothing but the excretions of plants and animals. Mother Nature is hopelessly incontinent. Yet nature rarely stinks. In fact, people go out into the countryside just to get a bit of fresh air. If it starts smelling like shit, you can be sure you're about to come upon a pasture or field. There's just too much manure out there these days. The plants can't handle it. In the Netherlands alone, with its small surface area, the crops have 70 billion kilos of manure to process per year. That's 28,000 Olympic swimming pools full. What the plants don't want anymore stays in the soil and contaminates the life there. Bacteria—nitrogen-fixing or not—give up the ghost, nematodes cash in their chips, earthworms turn pale and move on to better climes. Instead of the 500 earthworms per square metre that occupy good soil, artificially fertilised ground has only from five to fifty earthworms in the same space. Abandoned tunnels collapse, plant roots get stuck, the soil slams shut, yields decline. The government tries to contain the flood of manure. The result: the farmer takes a wife, with a manure quota. He ploughs even deeper. But his best ploughing team, the worms, are gone.

Artificial manure encourages cheating. If you want to keep on farming, you've got to stick to rules that have been honoured for centuries. If the amount of food you take from the ground is greater than the amount of shit you throw onto it, the crops will shrivel up until there's nothing left to eat. If the amount of shit is greater than the crops can process, you'll lose both your yields and your reserves. If the chemical fertiliser factories dig up all the phosphate and burn up the last of the oil in order to extract nitrogen from the air, the food cycle will come to a grinding halt. Even though it once was so easy to keep it well-lubricated. With shit.

That cycle runs on solar energy. It's not as fast as a cycle that runs on oil or steam. You can only harvest one or two crops a year from the sun. If you really can't wait until manure turns into fertiliser, fertiliser turns into plants, and plants turn into the food you eat and then shit out again, you can magically extract the energy from the shit in one fell swoop by setting it alight. That's what they do in India. Overpopulation and soil depletion there have led to an extreme shortage of wood for cooking. Manure, on the other hand, is available in plentiful supply thanks to the holy cows. As long as the cows are sacred they don't get slaughtered, and as long as they don't get slaughtered they shit. In India's warm climate the cow pats quickly dry into flammable slabs. While a cow in the West is a way of turning inedible grass into food, the cows in India turn incombustible grass into fuel. Only half of the 750 million tonnes of cow manure ends up on the fields; the other half goes up in smoke. This way India saves 100 million tonnes of wood or 50 million tonnes of coal each year.

In India, dried cow pats are peddled as if they were precious cakes. And indeed they are—not for eating, though, but as cooking fuel.

Cooking with manure is done mainly on the steppes, where nomadic tribes travel about with their cattle and there's little wood to be had. In Tibet they cook with yak or sheep manure, in Mongolia with the manure of oxen, in China with camel dung, and in Arabic countries with donkey droppings. In the modern West there's little demand for cookers that run on turds. As romantic as it sounds to let your pot simmer on the shit of your girlfriend, your dog or your guinea pig, it's hygienically unacceptable. Yet plenty of households do derive their energy from manure. Not by burning it, though. To burn something is to combine it with oxygen. The main product of this reaction is carbon dioxide, which we already have much too much of, increasing the greenhouse effect. When there's no oxygen the manure doesn't burn—it ferments, just like it does in our large intestine, with the help of anaerobic bacteria. This produces methane. Methane burns very well; it's the same stuff that's in natural gas. The manure from one cow provides enough energy for five households; the manure from all the cows in the Netherlands would keep the entire population going. Unfortunately, the yields do not always offset the costs. So why not cook with humanure? Most human manure ends up in the sewer. Purification usually takes place in the well-known open tanks, using lots of oxygen. Although the purification plants look so sophisticated, they don't do much more than what bacteria have been doing for millions of years: eat whatever there is to eat. All that's left is sewer sediment—fifteen kilos per person per year—which is burned as rubbish. The problem is that all that sewer water makes the shit too thin. In order to ferment it efficiently you would have to capture it in its dry state, close to the source. It's still not clear how that would be done technically, but we can be certain that the days of water sewerage are numbered. In the meantime, work is being done to derive energy from urine. The ammonia in our pee can be used to run fuel cells. If everyone in the Netherlands were to participate, we could produce 100,000 megawatt hours, enough for 30,000 households. That's no more than the production of a mid-sized windmill park. But yellow electricity has at least one advantage: it works when the wind isn't blowing.

To use shit effectively you would have to collect it, and that is the biggest obstacle. Our entire infrastructure is based on getting rid of our excrement as fast as we can. The main reason used to be because it's dirty, but later it was because it can make you sick. Yet for centuries faeces and urine were used medicinally. A monument to this is the *Dreck-Apotheke* (1699) of Christian Paullini, personal physician to the bishop of Münster, with the subtitle that speaks for itself: *Wie nemlich mit Koth und Urin fast alle, ja auch die schwerste, gifftigste Kranckheiten, und bezauberte Schäden vom Haupt bisz zum Füssen, inn- und ausserlich, glücklich curiret worden* (How in fact almost all diseases and cursed sores, from head to toe, internal and external, even the most serious and most toxic, can happily be cured with shit and urine). The success of scatotherapy is partly based on the idea that you have to fight evil with evil. That filth wins out over cleanliness is an age-old experience; only a filthier filth can beat it. And what's filthier than piss or shit? You could always find an animal who could shit you back to health. Pigeon shit helped against baldness, and peacock shit against epilepsy. The remedy for all complaints, of course, was *primus inter pares*: human dung. Not only did it restore you to your old self, according to doctor and natural scientist Martinus Houttuyn in his *Natuurlijke historie* (1761), but it even kept you young:

> A Lady of quite some Distinction, by using this Dung Water, maintained the loveliest Skin and the most beautiful Colour that one could ever wish for until a very advanced Age. Here is an explanation of how she procured a Supply of this water.
> This Lady had a young, healthy Servant, who relieved himself in a tin-plated Copper Basin, which had tight-fitting Lid. As soon as he had had his Bowel Movement, he rapidly covered the Basin so that no steam would escape, and, when the Young Man judged that everything had cooled, he carefully collected the liquid that clung

to the inside of the Lid, and put it in a Bottle, to be used during the Toilette of his Lady like a precious perfume. This Lady never failed to wash her Face and Hands with it every day, and by this Means she succeeded in maintaining her Beauty for as long as she lived.

Suspicion remained, however, although less in the countryside than in the city. That difference goes back to antiquity. At that time, the great physician Galen advised against prescribing faeces for medicinal purposes to city dwellers and other fancy folk; such popular remedies would be too strong for ladies from the city and for children. If you were travelling or hunting you had little choice, but by then you were already in the countryside, where the people were hardened and had 'the character of mules'.

Outside the cities no one ever made such a fuss about poo and shit and manure, even centuries later. They had an odour you could trust. They smelled like good business and healthy soil. It was actually rather pleasant and almost erotic, as Heinrich Heine testifies:

> *Eine bessre Wärme giebt*
> *Eine Kuhmagd, die verliebt*
> *Uns mit dicken Lippen küsst*
> *Und beträchtlich riecht nach Mist*
>
> The warmth a milkmaid gives
> Is better yet, her kiss
> With lips so full and young
> Amidst the smell of dung.

'This aroma is just what gives the romanticised rendezvous with the milkmaid its special cachet,' declares Florian Werner in *Die Kuh*. 'Although we may not find the smell of cow manure especially pleasant today, let alone titillating, we mustn't forget that this olfactory note was much more common in the nineteenth century than it is now. Even in the early twentieth century, the participants in a course for dairy product experts who were involved in a blind taste test didn't find the milk they sampled especially tasty until a pinch of cow manure had been stirred through it. The odour of animal excrement used to be regarded as salubrious, so much so that some sanatoriums had stables at their disposal where sickly city-dwellers could refresh themselves with the healthy smells of country life in order to speed their recovery.'

Whether it's a 'smell' or an 'odour', stench is stench. Even in the countryside. Shit repels far more than it attracts; a turd on your doorstep is rarely meant as a sign of welcome. When the first god-denying city-dwellers went beyond the metropolis to live in the strict Calvinist villages of the Netherlands, they could expect to find a load of manure delivered to their door. If it wasn't the stench then it certainly was the sign of rejection that sent them back to where they came from. But even in the city, manure was the best deterrent for the thorniest situations. Not something rustic like cow manure, of course, but the more cosmopolitan measures that were locally available, in the zoo. In *Message to the Rat King*, Harry Mulisch recalls how the wildest plans were hatched in Amsterdam to disrupt the marriage of Queen Beatrix to a German. His favourite, he writes, was the plan involving lion manure:

> Someone had once heard that horses *always* bolt whenever they smell lion manure, even though the horses had been tranquilised up to their eyeballs. So the plan was to collect the manure, with the help of the republican lion keepers in the city zoo, and to scatter it along the route to the Westerkerk. Thus the people of Amsterdam would be treated to the sight of a runaway golden coach with Beatrix and Claus inside, their arms thrown around each other's neck in terror, racing pell-mell down the Rokin, across the Munt, via Rembrandt Square to the Wibautstraat, and then, with bells all a-jingle, towards Motorway 1 and Germany—while we, in the meantime, would be back at Artis paying tribute to the Lion of the Netherlands.

The plan was called off, Mulisch writes, when they heard that the Golden Coach was going to be outfitted with disc brakes. According to student leader Ton Regtien, however, a respectable number of people did climb over the Artis fence in the middle of the night, but in the end no one dared enter the lions' cage. Whatever the reason was, nothing was ever proven and the question remained: does lion manure really work? Although little research has been conducted so far, the answer is certain: yes. Of course. The only weapon that a horse has recourse to against large predators is its fear. If a horse smells danger, it flees. A horse robbed of its only weapon through training is called a police horse. But the fear that a potential prey feels for predators is never completely gone. Voles avoid the faeces of foxes; cows and sheep won't touch any grass that smells of panther shit.

If the royal wedding had taken place a few years later, lion manure wouldn't have been so difficult to procure. By then you could just buy it as a way of keeping strange cats out of your garden. Except it didn't work. Cats had apparently never heard of lions, and the only one you'd chase away was yourself. Lion shit smelled even worse than cat piss.

American farmers knew how to have fun with manure as a deterrent back when you could still camp everywhere for free. If another horde of hippies or other urban slackers threatened to trample your land underfoot, you could simply start your fertilising a bit earlier and put down a heavier layer. At the same time, the CIA exploited shit's low approval rating in Vietnam. There they disguised radio transmitters as tiger turds and threw them from the air onto the Ho Chi Minh Trail in order to monitor enemy troop movements, certain that no one would touch the transmitters. Perhaps the CIA had come up with the idea by studying certain moths that are camouflaged as bird droppings, spatters and all, to keep from being eaten.

The ancient Egyptians found more inspiration in the dung beetle, who moves forward pushing its ball of dung. They understood the basic principle of all ecology: the world is round and is held together by shit.

Thanks to shit, everything is connected. Shit is of supreme importance, not as the supplier of ambergris or as a substrate for the cultivation of tulips but as the lubricant for the entire ecological mechanism. It is manure that holds nature together and closes the circle of life and death, of eating and being eaten, of parent and child, of past and future—over and over again. The world runs on shit, and Mother Nature knows it. That's why she never flushes her toilet.

Not when it comes to seas and oceans, that's for sure. Without shit there would be no life, without life no shit. It rains shit from the anuses of fish, lobsters and dolphins. The heavenly white sand of so many tropical beaches consists of little else but fish poo. Parrotfish nibble on the coral reefs there, and each year they add 100 kilos of beach nourishment per fish out of a torrent of indigestible little bits. In 2013 Dutch researchers found that reef sponges are even more indispensable, if that's possible. They convert the waste products of living reef coral into sponge poo, the basic food for countless snails, worms and crabs. But even the production of the sponges is a drop in the bucket compared with those trillions and trillions of plankton that treat the deeper levels of the sea to a rain of shit from the energy they've collected for them in the sea's uppermost regions. To estimate the value of the role of shit particles, marine ecologists tell us

> it's handy to see them as organisms; they consist partly of lumps of living cells, they eat and supply nutrients and organic material, and they serve as food for marine animals. So 'populations' of poo grains form a dynamic component of marine ecosystems, where they make a quantitatively substantial contribution to the energy flow and the food cycle.

For ordinary people, these heroic achievements of international marine excrement are far beyond their realm of experience. What does astonish them, though, is shit's popularity as a luxury consumer product. The most expensive coffee in the world, for example, is harvested as dung. Before they're roasted and ground, the beans of this 'kopi luwak' pass through the intestines of a civet. Civets are also known as a luwaks or toddy cats (*Paradoxurus hermaphroditus*), and what they're mainly interested in is the soft casing of the coffee cherry; they then shit out the pits, which we call coffee beans. Initially this resulted in poor people's coffee. Because the natives of the Dutch East Indies were prevented by their colonial masters from keeping any of the coffee cherries they picked, they searched for the excrement of the civets that ruled the roost in the plantations like starlings in a cherry orchard. Now it's Western snobs and gourmets who crave kopi luwak. There are enzymes in the luwak's gastrointestinal tract that remove the bitter proteins from the beans, which tends to refine the taste and increase the price considerably. A 250-gram pack sells for fifty euros—plus the price your conscience has to pay, since the civets are usually kept in captivity under miserable conditions.

The best coffee is a poo product. The taste of the coffee bean is perfected in the intestines of a luwak.

Before the chemical industry was born, shit and piss were useful raw materials. With shit you could soften leather, make paper or plaster walls. There was nothing like cow manure and straw for filling the spaces in half-timbered houses. After the emergence of chemistry, people considered themselves far above such messing with faeces and urine. Decked out in their white lab coats, chemists like to forget that an element like phosphorus was discovered in urine—and by an alchemist who was searching for the philosopher's stone. That was in 1669, when urine was still revered by alchemists as 'the golden stream'. When the stream was evaporated, a liquid appeared that glowed a mysterious green in the vapours of the retort. Phosphorus would continue to be obtained from urine for another century, until it was discovered that much more phosphorus could be found in bones. In our time, phosphorus is simply quarried from phosphate mines and squandered in the sewers. It is estimated that recoverable reserves will be depleted within a hundred years. Prices are already increasing at such a rate that recovering phosphate from urine and excrement is starting to pay off. We've come full circle.

Economically and ecologically less important, but no less charming, is the use of shit as a souvenir. You haven't been to Gunnedah, a town north of Sydney, until you've brought home a little sack of koala manure to remember the Koala Capital of the World by. In Alaska you have a choice between earrings, key chains, tie clips and—I kid you not—lip balm, all made from moose faeces. A turd as a memento, you can't get any more appropriate than that. After all, what is shit but the memory of whatever preceded it? That's a question that yields both philosophical and biological fruit. A biologist rarely steps in shit, out of sheer respect. What may be a source of irritation to someone else is a source of information to a biologist. He can deduce all sorts of things about the producer of the turd from its shape, consistency and orientation. That's why biologists usually keep their eyes glued to the ground. Only birdwatchers are willing to make the occasional slip-up.

When animals defecate, they leave long-lasting clues behind that can benefit ecological research. Take all the *chetae* found in fox poo, for instance. These are the living bristles by which an earthworm wriggles its way through the soil. On the basis of these *chetae* it was determined that more than a third of the total diet of foxes in the English countryside consists of earthworms. That's a clear call for modesty, for the fox as well as for the hunters who slander it with insults like 'chicken thief'. By using genetic techniques you can study a bear's excrement and find out what it's eaten, or whether it's under stress, or what its origins are. When technology proves inadequate, specially trained Labradors help create a profile of a bear's life without killing it. The dogs track down the bear's excrement, from which researchers are able to determine the animal's distribution. If rhinoceroses have to be moved for their own protection, it's helpful to send out their turds in advance to give the new territory a familiar smell. Researchers on the lookout for eagle shit are assisted by the elegant sunburst lichen (*Xanthoria elegans*) that grows on it. With its bright colours, this lichen makes it possible to see from great distances where the eagles foul their home rocks in the Arctic. In 2012 a new colony of emperor penguins consisting of 9000 individuals was found in Antarctica in the Princess Ragnhild Coast area. The animals gave themselves away by their excrement, which stood out clearly on the satellite images as brown against the white sea ice.

If you want to make use of shit, no matter how, you have to handle it. And whatever you handle is likely to contaminate you. So faeces researchers can count of being the butt of jokes, and cesspool cleaners are shunned. With shit on your shoes you're already an outcast. People who write books about shit are regarded with suspicion, and in India those involved in the large-scale collection of human faeces have traditionally been from the very lowest caste.

    I wouldn't know what else to call this other than disgraceful. Until I read about the lecture that Joseph Jenkins gave to six hundred nuns on the subject of his humanure toilet on Earth Day 1995. The nuns themselves had invited him. But why, in God's name? What did a group of otherworldly nuns have to do with something as worldly as shit? Was it a matter of ecological or religious redemption for them? After the lecture came the opportunity to ask the audience why they were interested in human manure, of all things? Because, they said, they were 'the Sisters of the Humility of Mary'. For them, defecation was an exercise in humility. As it is for all of us, right?

8

From One Anus to Another

Defecating is not a spectator sport. As delightful as it is to perform the act yourself, it's not much to look at. The grace of a gymnastics performance, the beauty of high diving and the spinning of a discus thrower are sadly absent from this form of physical exercise. Yet people watch it all the time. Among animals. The fact that you've never seen your neighbour or your old aunt in the act of shitting is only because they're human beings, too. Just as sport leaves trotting to horses and flying to pigeons, shitting has been outsourced to animals. Not to horses, unfortunately, with their majestic buttocks, or to cheerfully crapping bunnies, but to the only animal that humans have ever wanted to be: the dog. And although there's nothing exemplary about it from an aesthetic point of view, millions of people go outdoors three times a day to watch their dog shit. Passers-by and neighbours get to watch for free. Their reviews can be read in the local media, though little of it is appreciative. In terms of indignation, no war or disaster on the front page can match the lamentation about dog poo in the letters columns. How many millions of kilos it amounts to, how it stinks, how slippery it is, why no one does anything about it. To limit the damage and embarrassment, more and more dog owners are cleaning up after their pets. It's amazing to watch them do something that they would never dare do with their own turd: pack it up in a plastic bag, steaming and warm, and throw it away. While people in the past could act as if nothing ever came out of their dog, now they have to walk around with their little plastic bags, looking inside for traces of strawberries or tasty nuts. The telephone has now become mobile, and so has the turd; it takes some getting used to, but it's worth the trouble. Nice and tidy, says the dog owner to himself, but the dog is totally baffled. It looks at its owner with a kind of despair. 'I don't do that to you, do I?'

Owners watching their dogs shitting is something you see every day in the city. On the other hand, you never see a dog watching its owner shit. They're allowed to come right into the house with us, these best friends of ours, and into the kitchen. And many people even let them into their beds. But when it comes to the toilet it's here and no further; this is strictly a people room. Maybe it's just as well that a dog doesn't get to see what its owner is doing in there. It would break its heart: even the nicest turd is mercilessly flushed away. Gone.

For a dog, a turd isn't filth that urgently needs to be got rid of. It's a precious means of communication. The scents it gives off form words, sentences, announcements. For a dog, a turd is actually a message. Two turds are a tweet, three are a memo. A street full of turds is a library full of romance and information, with here and there a small but not insignificant little poo that's been artfully twisted into a proud whipped cream dessert and is understood by every dog that comes along as a poem in itself. A dog never steps in its own turd, or that of another dog; to do so would be deeply regrettable. For it, throwing a turd away is tantamount to book-burning. Without excrement, a dog cannot make itself understood. We gag them at their nether end. The poor despairing creatures sniff around for something to hold onto. The gentle whining dogs make even though they've just been let out is nothing but a cry for freedom of expression. It's a good thing there are still scent markings on trees and lampposts. You can't throw them away that easily.

Of the two parties involved in letting the dog out, the human is the most to be pitied. There the dog is in charge of the person. While it sniffs one bit of news after another, getting all wound up from the libidinous odours, adding its own gossip to the last with one lifted leg, its owner plods along beside him like a fool. He doesn't smell a thing. Like a blind man who can still make out vague contours, he does notice that a turd stinks, but the message escapes him. For a dog, a lamppost is all meaning, a source of knowledge, a beacon in the sea of ignorance that its master calls 'the world', but for a human it's no more than a way to make things out at night. To understand how such an animal feels in a street full of lampposts, take it to a library the next time you go for a walk. There the world is reversed. Now it's the dog who doesn't understand what its owner sees in all those books and magazines. Why does he laugh at that one back cover, and look around furtively before reaching for the next little book? Is there something to smell here? The dog may be an illiterate, but its owner is an anosmiac.

The dog is one up on the man—because it looks out of its nose. How on earth did this come about? It's because humans were once apes. Apes live in trees. They jump from trunk to trunk, from branch to branch. You don't need your nose much to do that. If you only look with your nose, you'll end up lying at the bottom of the tree in one or two jumps. What's needed is a good pair of eyes. Not eyes like those of an ordinary mammal—a cow, for example, on either side of its head—but eyes in the front, side by side, so you can properly gauge the distance of a jump. Stereoscopic. With eyes on each side you can see from every angle, which comes in very handy for a cow who has to watch out for wolves, or the steer. But such eyes are no good for measuring distances. That's why you don't see very many cows swinging through the treetops. Only two forward-looking eyes will allow you to see depth. But everything has its price. The migration of the two eyes to the front of the face left little room for a nose. The ancestor of the apes, the tupaya, still had a real snout, but by the time the anthropoid apes came along the snout was virtually gone. Humans have barely enough nose left over to prop their eyeglasses on. Our nose, like the appendix and the wisdom tooth, is a rudimentary organ—although for some it's less rudimentary than for others.

We're not the only ones who lost our noses in the trees. Birds can barely smell anything anymore. Apart from exceptions like the American vulture, birds don't sniff at their food until it's already been eaten, as a final check, and they do it via the nostrils in their beaks. Reading shit is also something birds rarely do. Instead they communicate with eye and ear, just like humans. That explains the good relationship between man and bird; the Society for the Protection of Birds has far more members than the society that protects our own class of mammals. Anosmiacs united.

Once we came down from the trees and went from ape to human we never did get our noses back. Evolution is a one-way street. Mammals that left the land to go back to the water in order to become whales never recovered the gills they once had possessed as fish. We have to make do without noses in a world full of expressive smells. The Dutch writer Annie M. G. Schmidt got it right with one of her first poems for the illustrious journalists' cabaret company *The Octopus*:

> Long ago we lived in the trees,
> I think I'd like to go back there, please.

You can see something of the wistfulness at the thought of a lost paradise in the joy of children who build huts in trees and the pain in the hearts of their parents when a giant tree is chopped down. Nor is it difficult to see the tree dweller in tram passengers hanging from their straps. But the tree dweller is most easily recognised by that silly nose. A human being is an ape without a tree. Too much earth beneath the feet. Too little smell in the nose.

You may have both feet on the ground, but your nose is still living in a tree. That tree is yourself. Ever since humans began walking upright their noses have been suspended more than a metre and a half above the ground. There's not too much to smell up there; it's like trying to read the newspaper a metre and a half away. The only way you can smell what your feet can't smell is with your nose on the ground. There the odours are as heavy as a blanket of mist hanging over a winter meadow, an internet of damp fragrances that benefit anyone who can smell them. Mice, dogs, insects and snakes are online there all day long; elephants, who, like us, are too tall, are connected by their trunks.

It's fun to watch a dog jump up to reach the height of its human, but it's just as instructive for you as a human to descend to the level of your dog. You'll never smell the world as keenly as your dog does. While our ancestors let their noses languish in the trees, the noses of the dog and the wolf became razor-sharp. The olfactory mucosa in our nasal cavity is no bigger than a postage stamp and is located high in our noses, while in a tracker dog every bit of snout that isn't needed for teeth is full of these postage stamps, up to more than fifty of them. In addition, the sensory neurons are much closer together. In the brain, where all stimuli are processed, the olfactory lobes in a dog are a hundred times larger than ours. All in all, a dog smells many things a thousand times better than we do.

That's not only to the dog's credit. It's also our own fault. We do too little with our noses, and as a result they waste away even further, so we use them even less, etc., etc. Use it or lose it. We've come to trust more and more in our eyes. There is still something to smell in an old book, but televisions and computers are completely geared to the eye and the ear. And no dog watches TV.

Humans have theatres where they see things and concert halls where they listen to things, but no smell theatres. A fun night out together for a good smell is not something we do. But our noses aren't really *that* bad. One drop of perfume can fill a whole shop with fragrance. To be able to smell violets you need 0.000000003 grams of aromatic substance per litre of air. To give you an idea of what this means, you'd have to make a hair-fine tube stretching from here to the moon and fill it with one gram of the aroma in question, ionone. One millimetre of that tube contains enough ionone to enable you to smell violets. A dog can manage with even less, but you'd never know it by its response; the smell of flowers doesn't do anything for a dog. Unfortunately, humans are born grouches. They're more likely to smell stench than violets. Fifty times less methyl mercaptan is needed for that, the sulphur compound found in the most disgusting farts. You can do much more than the basic things with such a skill, but you've got to learn how. Because we all smell high up in our noses, where a normal breath of air rarely penetrates, you have to learn to sniff—preferably repeatedly, because it's the alternation that makes you conscious of what's out there. After a while, really talented people can smell where a wine comes from, what its vintage is, and from what grapes it was made. But that's about as doggy as a human can get. At the very most a human can learn to read turds at the most basic canine level. What gets in the way of such a reading lesson is the immediate impression you get from a turd: its stench. When humans don't like the cover, they don't read the contents.

Most mammals read each other's excrement instead of a newspaper. But the news source is strictly regional. In the main article the depositor first claims the region for itself, which it regards as its own territory. Just like a real newspaper, the contents are only current for one day, maybe a few, and they have to be refreshed on a regular basis. For a large distribution area the report is also written in urine. Letters to the editor are sent by the kidneys and adrenal glands as well as by the sex glands. Lampposts, trees, large stones and other salient features of the landscape are marked with a drip here and a drop there. But there are other ways to do it. Antelopes smear branches from a scent gland near their eyes, dogs sometimes rub their entire body against a wall in order to stamp it with their body smell. Glands on either side of the anus add news and background information to every piece of excrement that passes through. But turds are always a special medium. They mark themselves; here, medium and message are actually one. Naturally it's more difficult to distribute your turds over a large surface area than the contents of your bladder, but animals like the hippopotamus have found a way to deal with this. They spin their tails around while they're shitting so the faeces are broadcast like manure from a manure spreader, making for a rather amusing sight in a zoo full of visitors with expensive coats.

Why does a male dog lift his leg to pee? Because he has no hands. When a human male pees against a tree, he uses his hands to take aim; a male dog does the same thing with his hind leg. Since leg and willie are connected in the groin, when the former is raised the latter follows suit. This requires the dog to curve his rump towards the tree, so that his head automatically turns away from the scene.

It's quite a performance. If seven hundred fifty thousand male dogs each marked a dozen Dutch trees every day, that would amount to almost ten million pools of pee all together. And all this effort would be in vain, since every territorial mark is peed over by the next dog as quickly as possible. Clearly something has gone wrong. Long ago, when dogs were still proper wolves, only the Great Leader of the pack would lift his leg, thereby underscoring his power in scent and colour. Our dogs no longer live in packs, and for them the Great Leader, oddly enough, is a human. Actually, it shouldn't be the dogs but their owners who go from tree to tree, lamppost to lamppost, leaving their marks. It would save the dogs a lot of work.

So why do dogs pee so high? Can't they aim a little lower, like on the pavement? They could, but the rain would soon wash the smell away. Not only that, but a dog's nose is at the same level as his willie; sender and receiver are coordinated. So aiming horizontally produces the best results. Wouldn't slanting upwards be even better? No. If the highest territorial mark were the best, then the biggest dog would always be the highest in rank. But dogs don't care who's the biggest. A dog doesn't know whether it's a Pekinese or a Great Dane; it still thinks it's a wolf. That's why little yappers bark so valiantly at big brutes; every dog, no matter how big or how small, is allocated the same portion of caninity.

The doggedness with which animals keep staking out their territory shows how important it is for them to have their own little domain. It's an extension of themselves beyond their bodies, essential to getting enough to eat and being able to offer something to a partner. To acquire a territory, animals readily sacrifice the highest good that we humans can imagine: freedom. Even a wild animal is imprisoned in hundreds of ways. A fish is restricted to its water, a squirrel doesn't dare come down from the trees, most insects are dependent on that one plant that they happen to relish. We know of birds who spend their entire lives in a small area of the tropical rainforest no bigger than a city block. Birds! Even migratory birds who cover thousands of kilometres a year aren't as free as they appear on TV. They travel fixed, narrow routes on a fixed schedule from A to B, and back again from B to A. These are not romantic vagabonds; they're wage-earning drivers longing for hearth and home.

Odours surrounding the territory are primarily meant to serve as boundary markers. Yet other individuals of the same species never recoil in terror and head for the hills. On the contrary, the smells entice them to come and take a leisurely sniff. The territorial marks and turds are not the animal version of an electric fence, but a notice containing information on accessibility. Like us humans, animals can't live without a sea of information—a network of informants, a kiosk, a town pump, an office water cooler. In lieu of paper, the information is printed on shit; in lieu of an alphabet it's presented in smells and colours.

Sometimes intruders are so hungry or sex-starved that they pay no attention to the warnings at the territory gate, but usually a great many fights are avoided with good information. This works to the advantage of both parties, since the pugnacious are always defeated sooner or later. And thereby shit and piss make a laudable contribution to peace and prosperity.

Even humans manage to use territories in order to keep the peace. Conflicts only break out when borders are not respected: war between countries, quarrels between neighbours. As inveterate visual animals, we draw our borders not by smell but by sight. Thanks to walls, fences, doors, nameplates and canals, we know how to find family members and friends in cities with hundreds of thousands of people living next to and on top of each other. Our homes are filled with personal items that guarantee a sense of safety; the first thing we do in a hotel room is spread out our belongings in order to claim the room as our own. Interestingly, the next thing we often do is go to the toilet. We act as if marking with excrement was something we had left behind ages ago, but the worst possible violation of your territory is somebody else's turd on your doorstep, your predecessor's faeces in your rented room. As if to remove any suspicion, many hotels place a paper band around the toilet seat in advance of your arrival. We can see how well the old smell system still works by the irritation of non-smokers who are forced to breathe second-hand smoke from someone else's cigarette. The smoker is violating their territory. The worst thing is that the smell continues to cling to them, as if they had been pissed on. And it's not mainly about health, as we see in the train, tram or lift, when perfume and aftershave are annoyingly imposed on you. This is something that women suffer from in particular. Because their sense of smell is keener than that of men, they use so little perfume that a man can barely detect it; but men immerse themselves in aftershave until they themselves can smell it, which for a woman is almost unbearable.

According to Jean-Jacques Rousseau, civilisation began with the invention of the fence. Unlike faeces, you don't have to refresh it every day. So fencing manufacturers do good business. Civilised people have doors with locks on them to ensconce themselves behind. The only way the outside world can intrude is through the mail slot. You can put special stickers on it in a modest effort to keep out unsolicited printed matter. Only posties can violate these stickers, their actions hidden by prudish metal flaps. We feel safe in our own territory behind our fences and doors—until a randy tom cat goes on the prowl and gives us a good spraying, making it clear whose house it really is. That makes people angry. And that's the nice thing about the territorial impulse: we only acknowledge its existence when our own territory is violated. Someone pisses through your mail slot. Red with rage, you open the door. There's a turd on the doorstep. The turd itself is easily cleared away, but the message remains: our existence is being challenged.

An animal turd works like a magnet: it can both attract and repel. While it keeps rivals at bay, it also attracts possible partners. A turd's at its best during the mating season. It's like a sign on the door; animals from the same species can read from a turd when a female cat or a bitch is on heat. In horny expectation, the tom cats and male dogs gather until the hour arrives. The entire garden, the entire street, all the trees in the neighbourhood are filled with the scent of promise.

Humans entice each other mainly by means of eye and ear. As man bait, a woman might put on a nice-looking skirt. If the enticement is successful, she'll take it off again. Music is played with lyrics that leave little to the imagination. Yet for something as primordial as sex, a primordial sense like the sense of smell does come in handy, even for humans. With a fragrant turd you don't even have to touch your beloved. But modern partners don't want to have anything to do with ordinary body odour at all. Before he goes out on his date, today's lover washes himself until there's nothing left to smell. To make sure he's free of all odour molecules, he shaves off the hair in which they might nestle, even in the most intimate regions. Only when he's less pungent than a marble slab or a strip of stainless steel does he apply his aftershave and lotions. Clouds of his artificial scent rise to meet her from a distance. There's not a trace of human aroma between them, neither in her smell nor in his. When vested for love, a human displays with the sexual fragrances of other species, animal or even vegetable. Girls flirt with the help of flowery perfume, as if they were out to nab a bumble bee.

As far as plants go, the leaf (verbena) or the bark (cinnamon) is sometimes used, but what parfumiers generally prefer are the reproductive organs, the flowers. Flowers were originally intended to attract insects. Humans and insects went their separate evolutionary ways hundreds of millions of years ago, but oddly enough they fall for the same fragrances. For example, both are wild about faecal aromas. The intoxicating smell of jasmine on a sultry evening is all owing to the presence of skatole, the same substance that makes shit smell like shit. But the fact that jasmine smells like shit is not to say that shit smells like jasmine. The art is in the dosage. To attract humans, you have to add so little skatole to the perfume that you notice it and but don't recognise it as the smell of poo. While a turd screams at you to stay away, jasmine whispers you into the kingdom of sensual pleasure, and with the same voice. What seemed so innocent and vegetable furtively awakens your deep animal passions.

As for animal substances, the better perfumes contain the sexiest juices of all the world's fauna: musk oil, civet semen and beaver semen. These are the substances by which musk deer, civets and beavers turn each other—and us—on. They constitute a kind of Esperanto, an inter-animal world language of voluptuousness. But here, too, you've got to be sneaky. To steal the heart of your lover, it's best not to burst into the house via the front door but to knock gently at the window. And you have to present your message in special wrappings. To have any effect, you add the more volatile, flowery fragrances as the top and middle notes; the animal fragrances form not the melody but the base notes of the aromatic music. Parfumiers have a word for this base note: booster, a substance that brings out the best in other fragrances without betraying its own origins, the way a pinch of salt can revive all the smells and tastes of a meal. But even a pinch can be very expensive; you need hundreds of thousands of flowers for a litre of jasmine oil. That's why skatole is often prepared synthetically.

During the sixties, the prospect of arousing a woman at a distance with the help of a fragrance gave researchers an idea. A similar experiment had just been carried out with insects. Female silkworm moths attracted male silkworm moths from kilometres away with a few molecules of a particular pheromone, bombykol. Irresistible. Imagine if such a thing were possible with humans! Expectations were great. Perfume manufacturers couldn't wait to come out with perfumes containing pheromones. Some even began production, but the effect was disappointing. A little cloud of perfume intensified the atmosphere on both sides, like beautiful music or sexy underpants, but there was no evidence of olfactory remote control.

Apparently our sense of smell just isn't good enough. And indeed, there's a lot about our noses to find fault with. But occurring at almost the same time as the pheromone experiment was an even more hopeful discovery: a second human nose! In addition to the familiar, not-so-great nose, it turns out that humans also have a vomeronasal organ, a nose within the nose. Animals were already known to possess such a thing. Cats use it when they smell catnip or munch on cat grass. They curl up their lips and stare cross-eyed into the distance as if they were stoned. This curling of the lips is called the flehmen response and is common among many animals. They throw back their heads as if they were stifling a loud laugh. The pressure of the lip muscles opens the entrance to the vomeronasal organ. What we learn from this silly behaviour is that the sense of smell is connected to the most arousing areas of the brain.

In humans, the vomeronasal organ is mainly associated with the embryo. It withers away before the baby is born, so it is regarded as rudimentary. But if you know where to look, you can still see it in most adults—or the entrance at least—as two indentations at the base of the nasal cavity on either side of the septum, a centimetre and a half from the nostril.

Does it still work? Women's magazines have known for years that it does, and scientific journals don't dare lag too far behind. Our second nose should be sensitive to pheromones, just as it is in animals. Pheromones are like hormones, except they don't do their work inside but outside the body. And like hormones, pheromones, in small quantities, have a major impact on your behaviour and mood. Women's magazines are always reporting that women become aroused by the smell of male armpits. This smell contains a pheromone, androstenol, which is also present in the saliva of randy male pigs. You get aroused just reading about it. Even more persistent is the report of female students who experience menstrual synchrony due to sleeping in the same room. This story has been going around since 1971, sometimes in exaggerated form as a report of menstruating nuns in far-flung convents, although it was difficult to corroborate experimentally. Statistically there was something fishy about the original research. Even if you don't have a head for mathematics you can figure out that two women who have their periods one week in four will find themselves in the same phase more often than you might think at first. They're never more than fourteen days out of sync, and there's an average of one week between the days when their periods begin. And if women really did respond to the smell of male armpits, and if the smell of male armpits really was the same as the smell of rutting male pigs, then pigsties would exert an enormous attraction on human females. *Farmer Wants a Wife* would be over in no time.

Scientific scepticism has never been able to harm the perfume industry. The only thing connecting science to perfume are the lab coats worn by the men in the cosmetics commercials. These men bottle pig slobber as an irresistible aroma. Now even hair gel contains pheromones. 'Attention: extreme attraction' warns 'Date Magnet' from the once-so-respectable Schwartzkopf. 'Men of the land: make women putty in your hand!' and 'They'll jump all over you!' Christine le Duc tells me encouragingly with her pheromone spray 'Lure'.

All this fire had already been predicted fifty years ago, and by the man who kindled it, no less: Alex Comfort, London gerontologist. While the first college girls were menstruating synchronically (or not) in 1971, he wrote in *Nature*:

> Science fiction has an awkward way of coming true. Nearly ten years back I wrote a science fiction novel which turned upon the discovery, use and abuse of human pheromones. A pheromone is a substance secreted by one individual which affects the behaviour of another—an olfactory hormone. In the story there were two—3-blindmicyn, which was the universal aphrodisiac, and cocuficin, which excited hostility between males. The main aim of the book was political satire, but the biology unfortunately took over—only biochemists and psychoanalysts found it funny, though the girl who typed it took it for a factual history of my life.

Pheromones seemed too good to be true. Even researchers in official university laboratories dreamed of substances engineered to entice men or to spray women into total submission. Others did their best to set them straight. These 'Stink Wars' have not yet subsided, but as the smoke of battle clears, the contours of reality become more visible. Any proof of the existence of mammalian pheromones has been found to be flimsy indeed.

But even if they do exist, they don't amount to more than a gentle nudge for the mammals that notice them. No directives from on high here, not even in a dog turd. Rarely, if ever, do you see a dog sniff another dog's turd and then do a one-eighty because its orders had suddenly been changed. Nor do humans let themselves be controlled by a turd's contents. Even if we had the most sophisticated equipment at our disposal to decode the message contained in a dog turd, we wouldn't understand it. It's no more couched in monosyllabic pronouncements than our newspapers and books are. Just as the smell of a cup of pea soup or a leather handbag is made up of thousands of components, some that augment each other, others that contradict each other, most of them irrelevant, yet others that set the tone from the background, so the smell of dog faeces is a complex interactive conglomerate. In order to enjoy all that information and to impart meaning to it amidst the rest of life's daily affairs, you have to have a dog's brains as well as its nose, complete with the worldview stored there. The idea that one smell has one meaning is just as outmoded as the idea that one instinct moves you to carry out one particular action, or that one gene is responsible for that one horrible character trait.

It's this inaccessibility that arouses curiosity about the contents of an animal turd. What exactly does the depositor want to tell the recipient? Confidentiality of the mails isn't the issue here; it's how do you get the damned envelope open? Is your dog gossiping about you in secret while you're just standing there? Are the dogs of the neighbourhood laughing at you? Or are they really as loyal as they claim? How delightful it would be if you could clean out the cat's litter box and learn all the latest juicy chitchat from the back garden, the way you get to hear everything about school when your child comes home. Or if you could get caught up on the latest news from a lamppost while taking the dog for a walk. Having a good chinwag with your hamster would certainly brighten up your mood—and your hamster's, too.

~

If you really loved your pet you'd take a course in turd reading, which is no more ridiculous than a course in palm reading or iridology. It seems to me that learning animal language before you venture into the big wide animal world is simply a matter of common decency. Bird guides would show you the way. Besides the birdcall they'd teach you the mating call, the warning cry and the contact twitter for every species. But such instructions would be no more useful to you than the sample sentences in *How Do I Say That in Estonian?* And whatever you might learn about the smell of their droppings would be even less useful. Mammal guides sometimes describe a turd in general terms (sweet, sickly, sharp), but that doesn't tell you what's being communicated. Humans fail their animal language exam before they even start. It's enough to drive you mad. We deciphered Sanskrit and hieroglyphics with a bit of puzzling, and we found out what the Chinese mandarins had to say to each other a thousand years ago, but what Buster thinks of Fluffy is completely beyond us.

Maybe we should approach the thing from the other end. The fact that we're too stupid to learn animal language doesn't rule out the possibility that animals can get the hang of ours. Vocal cords not required. It's mainly a question of good communication skills.

The best chance you have for a good conversation is with your own family. Of all the animals, the apes are our closest relatives. To teach them our language you have to get them when they're young, just like human children. In 1931 the American psychologist Kellogg—no relation to the cornflakes magnate—and his wife raised a young chimpanzee, Gua, along with their little son Donald. But the vocabulary never went much beyond 'papa' and 'mama'. It was another American couple, the Gardners, who grasped the problem: chimpanzees simply have no larynx to enable them to talk. Sign language would work better. Their chimpanzee, Washoe, made it to 130 words, very intelligently strung together into little sentences like 'Give me sweet, hurry' and 'Who good? Good me'. Expectations were high. At the end of the last century, scientists were obsessed with the possibility of ape speech. We were this close to bridging the gap between human and animal. 'This close' is now long past, and we've pretty much said all there is to say on the subject. The apes have vanished into the zoos, the jungle or the vivisection labs. After a brief introductory interview they were rejected as too lightweight, banished from the human realm and sent back millions of years to the kingdom of the apes.

Here and there you may still find a researcher plodding away with talking apes. But language lessons for snout animals like the dog have lost all their appeal. Nose and ear are not on speaking terms. Ear does not understand nose, a situation that exists not only between humans and animals but even within our own species. Of all the thousands of smells that we can distinguish in spite of everything, only a few have names, such as 'musty', 'penetrating' and 'rotten'. And most of these are borrowed from another sense. Smells are called 'sweet', 'sour' or bitter', but essentially those are tastes. Warm and heavy smells are named after tactile stimuli, and 'clear' or 'dark' are more likely to describe something seen by the eye rather than smelled by the nose. Charles Baudelaire spoke of 'fragrances as fresh as a child's body, soft as an oboe, green as the greenest meadow'. 'I really no longer knew,' wrote Guy de Maupassant, 'if I were breathing music, or hearing smells, or sleeping among the stars'. In *The Name, The Nose*, the writer Italo Calvino sends his decadent main character to the perfumery of Madame Odile to find out what fragrance had been worn by the lady, now vanished, with whom he had danced at a masked ball.

> What I required of Madame Odile's specific experience was precisely this: to give a name to an olfactory sensation I could neither forget nor hold in my memory without its slowly fading.

Usually we make do with a comparison: it smells like coffee, like sweaty feet, or like a first day of spring. This doesn't get you very far if you're a coffee roaster or wine connoisseur. Or a detective. According to Sherlock Holmes in *The Hound of the Baskervilles*, a fully qualified private eye should be able to identify seventy-seven different smells. Even Linnaeus, nature's bookkeeper, didn't manage that. In his *Odores medicamentorum* of 1752 he identified seven different classes, 'hedonically' ranked from pleasant to foul:

> Aromaticos: as aromatic as lavender
> Fragrantes: as fragrant as jasmine
> Ambrosiacos: as ambrosial as musk
> Alliaceos: as sharp as garlic
> Hircinus: as goat-like as sweat
> Tetros: as foul as coriander
> Nauseosos: as nauseating as faeces

Every attempt to find the right word for a new smell poses problems for the man of letters, as if his search had brought him face to face with a closed door. You're certain that the word must be tucked away in the cabinet behind that door, but you can't get to it. The word is on the tip of your tongue, but that's no help. It only shows us once again how much smell and reason belong to two different worlds. Ascribing a particular word to a particular scent is like trying to force a square peg into a round hole or drive a nail in with a screwdriver. Signals from the nose end up in the old vaults of the brain, the limbic system, which dates from a pre-verbal era. Words are processed in the neocortex, which has burgeoned like a block of flats across the brain's Old Town. Of course there are connections—in the brain everything is linked to everything—but the first thing a smell sensation evokes is not an idea but an emotion.

Thinking and feeling not only have separate headquarters but they also have separate transport systems. While reason prefers to give its orders electronically via high-speed nerve bundles in a way that is unambiguous and brisk, feelings trickle along chemically with the help of hormones. In order to deal someone a well-directed wallop, a great many nerves in the head, arm, and fist come into play, but the inclination to deal the wallop in the first place comes from your hormones. As every novelist knows, it's the main character's feelings that power their actions, not their well-considered decisions. Smells play an important role here. In Patrick Süskind's *Perfume* (1985), fragrance and feelings become completely merged.

> For scent was a brother of breath. Together with breath it entered human beings, who could not defend themselves against it, not if they wanted to live. And scent entered into their very core, went directly to their hearts, and decided for good and all between affection and contempt, disgust and lust, love and hate. He who ruled scent ruled the hearts of men.

Like all odours, the first thing the odour of a turd does is to release emotions. There are no neurons in a turd to make detailed announcements and give instructions. If a turd speaks a language, it's more in the form of poetry than prose, more advertisement than stock market report, more *Hello!* than *Le Monde*, sent by the heart rather than the head. If you're an animal, a turd can't teach you how to build a flying machine or who Charlemagne was, but it can help you pour out your heart, gossip, get into fights and settle them. A turd is a perfumed letter from heart to heart. It's a shame we humans, with our toy noses, can't use them to read what turds have to say, but maybe it's just as well. Those who listen to gossip run a serious risk of ending up being talked about themselves. Considering the way humans treat animals, I have no illusions about their view of us. The steam that rises from their shit is the truth about us that gets passed around. Although our noses don't pick up the interesting stories, the core of the message comes through loud and clear in the overpowering impact of the stench. The penetrating smell of unseemliness drowns out any nuance.

~

One good way of setting your neighbour straight is by sending him a turd. That'll teach him. The first victim, however, is always yourself. How do you pick up a piece of shit? How do you get it into an envelope? Fortunately you don't have to. The phrase 'piece of shit' alone says it all. Cursing! Swearing! Words with lots of shit in them derive their punch straight from your entrails, somewhere between your heart and your liver. Cursing is the opposite of a friendly greeting, swearing the opposite of prayer. You can pray for someone's wellbeing with great intensity, but you can just as wholeheartedly wish them ill. Most of the time that's about as far as it goes. You'd be downright astonished if the other person actually did drop dead or go fuck themselves. And yet cursing helps. I for one don't know how anybody could put an IKEA table together without cursing. It's a much more effective adhesive than glue or screws. And then just pray the thing doesn't collapse.

To add power to your cursing, it is recommended that you violate a taboo. The typical Dutch method is to curse the other person with some horrible disease or condition. There's also a lot to be said for a more general violation, towards God. Since it's the opposite of prayer, every curse is a form of blasphemy. But you can also go straight to the source. 'God damn it' —*Godverdomme*—is something you hear more often in Dutch than 'get tuberculosis!' (a real Dutch curse) or 'suffer from consumption!'; in Scandinavia people invoke the devil with '*Fy fan*'. For the unbeliever there's a broad range of taboos to choose from in the realm of sex. In America, 'fuck!' is not only one of the most popular curses, but if you can believe the films from that country it's also the most frequently used word in the English language. With the exception of 'shit!' In English as well as in German and French, the anus is used more often for swearing than for shitting. An interesting indication of the national preference for either shit or disease is the proper Dutch translation of the English phrase 'the shit's gonna hit the fan!': '*nou breekt de pleuris uit!*'—'now there's gonna be an outbreak of pleurisy!'

'*Merdre*' was the first word of the play *Ubu Roi*, which premiered on 10 December 1896 in the Théâtre de l'Oeuvre in Paris. Among the audience the shit almost did hit the fan. 'Despite the late hour,' one of the reviewers wrote after the performance, 'I went home and hosed myself down.' And so the avant-garde theatre was born. In France, that is. The Germans were already used to this sort of thing. In 1773 Johann Goethe had written *Götz von Berlichingen*—the name alone is now tantamount to profanity. In this drama, the brave knight Götz finds himself surrounded by his enemies. They call to him to surrender. 'Tell your captain,' Götz answers, 'that he can lick my arse!' This quip is still cited in Germany today, no matter what the circumstances. An inappropriate example took place during the Nuremberg Trials, when Göring slammed his fist on the table with the words, 'God damn it, I wish we all had the courage to limit our defence to three simple words: lick my arse! Götz was the first one who said it, and I will be the last!'

Hollering 'shit' has the same effect as shitting itself: out comes the filth, you're free of it, now it's someone else's problem. Very few things clear the air as capably as a good shitful curse. It's as if the brain's traditional linguistic centre had finally given way under the emotional pressure built up by the fetid air in the olfactory bulb. Before you know it, the words escape your lips like a fart from your arse: *Shit! Scheisse! Merde!* No shit, Sherlock.

Shit, Scheisse, merde: even the most common word for poo sounds like a curse. So what term should you use if you just want to talk about it? To prevent you from starting to swear before you've even begun having a serious conversation, a whole series of new euphemisms have been invented over the years. Faeces, excrement, stool—none of them help very much. There'll always be something fishy about shit. Due to the absence of a neutral word, the subject has simply become unmentionable in polite company. To get around these difficulties, people tend to talk about the place where the act is performed rather than the product or the function. Thus one goes to the toilet—or, as respectable people put it, to the loo. It's reminiscent of the way sex was dealt with until fifty years ago. In the absence of a nice word for fucking (the act), people went to bed with each other (the place). Today, as then, there are three social refuges where you can freely talk about shit, and each refuge has in its own jargon: medical, vulgar and infantile.

Doctors revert to the trick by which they were able to elevate their profession above the level of the common man for so long, the same trick the Catholic church used so successfully: the use of dead languages. There are no obscenities in a dead language. Doctors don't fuck their brains out in Latin and Greek, they cohabit; they don't shit, they defecate. Actually, faeces is just an ancient word for muck and filth, but everyone seems to have forgotten that. Even more serious is the tendency to associate medical terminology with sickness and death. The fact that doctors only deal with sick people is having its revenge. They have absolutely no knowledge of what's going on inside a healthy body.

Popular speech calms the fear of death and decay, of piss and pus, with humour. In popular speech, people seldom defecate or urinate; it's always taking a crap or taking a leak, to say nothing of pinching a loaf or draining the main vein. In folk tales there's no end of sliding on brown sludge, chamber pots being emptied on people's heads, exits used as entrances, and entrances as exits—all of it accompanied by snorting and sniggering. While it may not be everyone's cup of tea, real folk humour always calls a spade a spade. As the tram conductor said after one of the passengers had farted: 'Everybody take a whiff and it'll be gone in no time.' And this well-intentioned complaint from the Brabant farmer: 'No sooner do you settle down for a really good shit than you run out of the stuff.'

For children, shit is a gold mine. First there's the material itself followed by the words to describe it, and this goes on until they're ready for sex. For parents, children are a welcome outlet, giving them a chance to join in the poo talk without holding back. At the same time, however, the kids are already practising the use of common euphemisms. The Dutch writer Godfried Bomans provides a splendid example of this in his 'earliest memories' of *Pieter Bas*: 'Anna, who "had been in service to the master himself," insisted that all four of us take our turn on the chamber pot, for, she said, "a true Christian could not sleep until all the 'bad stuff' was out."'

> So four chamber pots were lined up against the wall. The Bas brothers sat themselves down and looked at each other intently, since the whole idea was to see who got the bad stuff out first. The first one to do so would climb into bed triumphantly and wait for the next winner. Together they would egg on the two stragglers, urging them 'not to let down the team', and these in turn would stare at each other with red faces, determined to give it their all. Poor Jozef! He always came in last. And when one after another had crept into bed and he was left alone on the floor, groaning, I sometimes felt sorry for him. So I'd straighten myself up and holler, 'How's it going?'
>
> 'Almost there,' would come the serious reply.
>
> Good boy, Jozef! But every now and then he'd be rewarded for his efforts.
>
> 'Everybody come and look,' he'd shout, very excited.
>
> And the three of us would fly out of bed, since 'what Jozef did was nothing to be sneezed at'. We'd study the results with the eyes of experts who knew a prize when they saw one and tell him it was a 'whopper', while the happy owner would receive the tribute with a grateful little laugh.

Unfortunately, because of its popularity among children, pooing has acquired a somewhat childish reputation. Ask for a book about poo in your local bookshop and you'll be directed to the children's department. There you can find a book that is called *The Story of the Little Mole who*

Knew it Was None of his Business. It's a story about a little mole who wants to know who shat on his head, which is actually the title of the book's Dutch translation. For an adult such a title is unthinkable; the word 'poo' alone is taboo. Too childish. So as the writer of a serious book about poo you're stuck with a vocabulary full of holes. Writing a book on a subject for which there is no decent word is like baking a cake without sugar, building a house without nails, or sitting on the toilet without having to go. How is this supposed to end? Back when it was all about sex in books, it seemed like a question of attrition. Sex book after sex book wore away the sharpest edges of words like 'cock' and 'cunt', but between 'cock' or 'cunt' here, 'penis' or 'vagina' there, and 'willie' or 'pussy' somewhere else, the gaps still yawned. In the case of sex these were filled in with provocative photos, erotic films and exciting lingerie. If you didn't have a word for it, there were always pictures. But it's trickier with shitting and peeing.

If you want to write about shit you have to make do with what the language has to offer. Nothing for it but to change register from time to time, be prepared to put up with ambiguities (it's a subject that can't help being funny) and enjoy childish pleasures—and if the entire vocabulary lets you down, vent your feelings just like everyone else with a loud 'Shit!'

9

Water and Gas

Do plants shit? Does a geranium have to go every once in a while? They're suspiciously quiet, the plants on your window sill, as if they were actually doing something. But they don't shit. That's what makes them so popular as housemates. Just imagine having to take your begonias and African violets out for a walk along with your dog. Out of the question. Throughout their lives, plants retain all the excrement they produce in vesicles or crystals. These little rubbish bags are what give spiced food its pungent flavour. You season your food with plant shit.

So you don't have to walk a plant, but you do have to give it water. In a natural setting plants get water automatically from the clouds. But there are no clouds indoors; for a house plant, a living room is as barren as a desert. So plants have no business being in your house. If you still want a couple on your windowsill you'll have to provide the rain yourself, locally, with a watering can. This is not an insurmountable problem.

We're only too happy to give drink to the thirsty; it's a real work of mercy, and it's much less laborious than feeding the hungry, visiting the sick or burying the dead. But it doesn't help much. The next day the plants are dry again. What did they do with their water? If they had drunk it all up your African violet would be an African bush after a week and an African tree after a year. In actual fact, your plants look at you reproachfully from their dry saucers after just one day, manipulating your guilt feelings by hanging there limply, pure blackmail, with yellow leaves in reserve as a last weapon.

Do plants pee? No; full saucers dry up, but empty saucers never fill by themselves. A plant has no willie, no kidneys, no bladder. Plants don't pee, they sweat. They do it all over their bodies but especially from their leaves, through millions of openings, the stomata. The evaporation of all those drops of sweat provides the power to draw the water up out of the roots. Plants don't have to get rid of water by peeing; they drink water in order to sweat. They drink with their roots, using the water they absorb to pump up their cells like the air in bicycle tyres. If there's a water shortage the cells slacken and the leaves hang limp.

Humans pee. Boy, do they ever! A litre and a half a day is only the average. That's more than enough to step away from the subject of shit for a moment. The urine reservoir, the bladder, fills up several times a day and has to be drained. Whatever goes in by drinking has to come out by peeing. Yet something isn't quite right. A person drinks an average of three litres a day. That's twice as much as they pee out. A litre and a half has gone missing. And even more than that. If you count the liquid in your food that's half a litre more, and a quarter of a litre is released when you digest your food. Where does all that water go? There's only one explanation: you're leaky. As leaky as a plant.

So where's the leak? To find the leak in a bicycle tyre you immerse it in a bucket of water; for a leak in a human being you place the human in front of a hot stove and water will come out drop by drop. The person is sweating, just like a plant: out of lots of little holes. A human has two million sweat glands. That's more than most animals have. Cats, for instance, sweat only on the soles of their paws; if you look closely you can see the damp tracks where they've walked across the piano.

Humans don't sweat to pump up water, the way plants do. They have hearts to do that; the heart maintains a stream of red water to take care of internal logistics. Humans sweat to cool off. Other mammals have little use for such a cooling system because of their fur, which interferes with evaporation. But for naked apes in sunny Africa it was a godsend. By sweating they were able to endure the heat longer than their prey—as long as there was enough water on hand to replenish their reserves. In the tropics you need as much as ten litres of water a day, and in many places in Africa it's just not available. That's why Africans are so lazy—in the eyes of many old colonialists. A black person would rather doze against a palm tree, they think, than climb up to pick bananas and open a fruit and vegetable stand. What you need are white or yellow people, the ones who built the big cities and civilisations, from Babylon to Beijing and from Rome to Amsterdam. But white and yellow people come from cool lands, and black people come from the tropics. In the tropics you can easily become overheated. The best remedy is not to work too hard. Just lie there until the bananas fall of their own accord. If the colonialists had bothered to roll up their sleeves and get cracking, they would have discovered this for themselves. But they had one advantage: they had salt. Natural salt was scarce in the middle of Africa, far from the sea, and it was difficult to replace the salt you lost through sweating. There was salt to be had elsewhere, but no potable water. Salt had been found in the desert, but the people who found it let others do all the work. Camels. Camels transport the water from oasis to oasis—outside and inside. A camel can drink 100 litres of water at a time. Where do they put it all? Not in a stomach; that's too small. And not in a hump; that's full of fat. If a hump shrinks after a long journey, the camel restores it by eating, not by drinking. The water it drinks disappears into the tangle of its intestines, from which it's distributed throughout its entire body (now suddenly 100 kilograms heavier). But a camel is leaky, too. When it does a heavy job of work, it can lose ten litres of sweat a day. To protect its water supply, it pees only one scant litre from its enormous body, and even its dung is on the dry side. This enables the camel to stick it out for more than a week, working hard and without drinking.

Excess water that isn't lost through sweating or breathing has to be peed out. For an intriguing experience, try peeing and drinking a jug of water at the same time. It takes surprisingly little imagination to feel as if the water you had poured in at the top had run right out again at the bottom, something like the vague tingling sensation you get in your head when something goes in one ear and comes out the other. That's the work of your brain. If it sees two things happening at the same time, it assumes that a causal relationship exists between them. This is called understanding. But it doesn't always correspond with the facts. The water you drink now doesn't reappear when you pee until much later, which is a good thing. Your body needs time to extract useful substances from your drinks: alcohol if you want to be cheerful, vitamins if you want to be healthy. This happens mainly in the gastrointestinal tract, which also loses water with every turd you produce, but not much. Unlike peeing, pooing is responsible for keeping water in; most of the water you drink goes into your body via the large intestine and the blood. While shit has never really been in your body at all, urine has been to every nook and cranny. Water and blood together travel to the kidneys. These are two bean-shaped filters, as big as a hand, on either side of the spinal column. Here 99 per cent of all the incoming blood is purified and put back into circulation. Of all the water that flows through your kidneys, you only pee out 1 per cent. The exact amount is not regulated by the volume of water but by the salt content.

The main purpose of peeing is to maintain the salinity of your blood. The water level then takes care of itself, since salt retains water. A spilled heap of salt on the kitchen table spontaneously absorbs moisture from the air—that's how much it wants to be dissolved in water. You lug around fifty litres of water in your body—five bucketfuls—and you need half a kilogram of salt to retain it. Your blood is an inland sea of salt water, a legacy from the time when our distant ancestors were sea-dwellers. But our blood isn't as salty as the sea. With nine grams per litre instead of thirty-five, blood is only a quarter as salty. The water in your inland sea is brackish. If it gets too sweet you have to eat more salt or pass water until the salt content returns to nine grams per litre. If that doesn't happen, your body cells (which are brackish, too) will absorb so much water that they literally burst. This rarely happens in the Western world, with its vast assortment of salty food. We are more likely to suffer from an overdose of salt. To get rid of it, our kidneys have to work overtime or we have to flush them with additional fresh water. You get thirsty. So drink a lot and pee a lot, as long as the potable water is fresh. Otherwise you die of thirst, like a shipwrecked sailor in the middle of the ocean. Even though he bobs around on his raft in a million cubic kilometres of water, he'll die of thirst in the end. If he doesn't drink he'll die; if he drinks seawater he'll die, too. For the kidneys to pee out all the salt in one litre of seawater they need two litres of fresh. Seawater just gives the kidneys more salt to process, and the body cells are soon as pickled as the cells in a salted herring. The full-strength seawater draws the water out of the brackish cells, which shrivel up and fall apart.

Urinating is not meant to get the water out of your body. If that were true, you might just as well drink less water. That makes sense for a land animal, since potable water is scarce on dry land. Water in your body serves as a means of transportation for substances you have to get rid of. Your urinary passages are your own personal sewer, and a sewer doesn't work without water. Besides salt, water also drains off the products of metabolic decomposition. Carbohydrates and fats present few problems during combustion. They consist mainly of carbon, oxygen and hydrogen, which can be burned to form water and carbon dioxide. Just like plants, we can exhale them without giving anyone offence. If we were to live from carbohydrates and fats alone we wouldn't need the LADIES and GENTS anymore; the range of our cars would ultimately be determined by their own tank capacity and not by the capacity of our bladders. Unfortunately we eat the wrong things. There's too much protein in our food. Protein contains nitrogen, and nitrogen isn't as easy as oxygen, carbon, and hydrogen to process into something harmless. The simplest possibility, ammonia (NH_3), is highly toxic. If you're a marine animal you can rid yourself of nitrogen in the vast oceans of the world, but on land you have little other choice than to firmly incorporate nitrogen into the more complex urea ($CO(NH_2)_2$), which has to be peed out with lots of water. Urea constitutes half of all the dissolved substances present in urine; the other half consists of more than a hundred chemicals.

Urine is yellow. It's usually light yellow, but the more concentrated it is the darker it gets. Vitamin B2 makes the yellow nice and bright. If you'd like something other than yellow, you can colour your urine by adding rhubarb or senna (brown to pink), beets (red) or blackberries to your food. You can give it a pleasant odour by inhaling turpentine, which makes your urine smell like violets or roses. Fashionable ladies in ancient Rome did this with abandon, although it isn't healthy. Better to eat asparagus, valerian or leeks, each of which gives urine its own pronounced aroma. But you can also do nothing at all. On its own, fresh urine has a delightful beef bouillon smell, especially if you condense it. The salty taste closely corresponds to that of bouillon, although it's too bitter for most people.

Urine, with all the substances it contains, is supersaturated, so it easily precipitates to form grit or stones. As long as these things are in the kidneys you don't notice them very much, but as they move towards the exit they naturally run into obstacles and get stuck as kidney stones, renal pelvis stones, ureter stones or bladder stones, which can be very painful. As people eat more meat and become less active they're more likely to develop such stones. In the Netherlands the count so far is five (women) to eight (men) out of 10,000 individuals. Fortunately the range of treatment possibilities is steadily increasing. With a lithotripter, many stones can be rendered harmless externally.

Almost all the dissolved substances disappear into the sewer with our urine, which is a shame. That's a loss of precious materials. Many Westerners take multivitamin pills every morning, which means that the Dutch, the Americans and the Germans have the most expensive urine in the world; very little of the vitamin content they take is actually absorbed into the body. Water purifiers complain about the high proportion of medicines in their sewer water, which ends up treating entire population groups without a doctor's prescription, after purification. Oral contraceptives interfere with the reproduction of frogs via the sewer. Conservationists fear the worst for the survival of marine animals.

Instead of whining about it, environmental experts would do well to take an example from our forefathers, who saw their urine for what it is: a rich source of chemicals. The Romans used it for washing. The ammonia that was released from the urine after a few days of rotting combined with the grease from the laundry to form a liquid soap. Laundry and urine were collected by the same group of workers. From every corner of the cities of Rome, these *fullones* (fullers) picked up the urine from the local residents in large amphorae to be used in the laundries, which were understandably located very far away. The dirty togas and stolas were placed in stone troughs and soaked in urine, then trampled like grapes. Finally the urine was rinsed out in a stream or river and the laundry could be bleached with sulphur on a sunny field. Filthy as it was, the work was well paid. Emperor Titus Flavius Vespasian (69–79) became aware of this and levied a high tax on it. When the fullers complained to the emperor and referred to their malodorous working conditions, the emperor uttered the immortal words '*pecunia non olet*'—money doesn't stink. Urine was also used in the dying of textiles. The ammonia made the leaf- or root-based pigments water soluble so they could be absorbed by the textile fibres. Once they had been saturated with the dye, the lengths of cloth were hung in the fresh air. There the oxygen reversed the chemical process and the pigments became insoluble once again. Otherwise the fabric would fade in the first washing. Urine was also indispensable in the fulling of wool, until the emergence of the chemical industry. The ammonia in the urine caused the wool to become matted, which thickened the cloth.

Urine was famous for its healing powers even before it became laden with vitamins and other medicines. And urine therapy still has its adherents. Every morning they drink a glass of their own water. Christians feel encouraged in this by the biblical injunction 'Drink waters out of thine own cistern, and running waters out of thine own well.' Hindus have traditionally regarded their urine as *shivambu*, 'the water of Shiva'. Among the modern practitioners of urine therapy were Mahatma Gandhi, John Lennon, Jim Morrison and the Indian prime minister Moraji Desai. It can't do you much harm: the worst that can happen from drinking your own urine is getting your own illnesses, and you've already had them. There's no bacteria in urine fresh from the ureter. But for centuries the use of urine as a diagnostic tool was even more widespread than its use as a medicine. If you got sick during the Middle Ages, you'd fill a bottle with urine first thing in the morning and carry it in a wicker basket to the uroscopist. He would hold the uroscopy flask up to the sun, study the colour, smell it, taste it, stir the sediment and listen to the bubbles before pronouncing his diagnosis. Sometimes he was right. If you had diabetes, for example, he could always taste it because the urine would be sweet; otherwise it wasn't. The ancient Greeks had an even simpler method. They poured the pee over a stone and waited until it dried. If it attracted bees, you had diabetes. There are still doctors today who simply stick a finger in the suspicious urine and lick it. But most of them now use a dipstick, by which they can determine the proportion of acid, protein and blood. Amino acids like leucine and tyrosine indicate liver disease. Bacteria are cultivated in order to detect an infection. Highly refined analytical techniques have been established to trace the remains of drugs in the urine of athletes or to narrow down the use of narcotics to weeks after the last use. Doctors haven't been this confident since the days of humoralism. Back then, health was a question of balance among the four bodily fluids: blood, phlegm, yellow bile and black bile. The best way to determine this balance was by examining the colour of the urine. If it was too yellow, that meant there was too much yellow bile, and the patient's temperament would be choleric; black urine indicated malaria (blackwater fever); and red urine was due to an excess of blood. So you knew what you had. Then all you had to do was cure it.

The inability to urinate is very serious. In its extreme form it can kill you, in its mild form it can be very uncomfortable, even if for no other

Peeing is infectious and creates a sense of warm intimacy, even between different species.

reason than missing the relief you get from taking a good piss. There's little that can beat the delight afforded by a good hearty stream aimed against a tree, the liberating spatter in the toilet, the catharsis that comes after the thigh-squeezing anxiety of waiting your turn. No one has expressed this better than the French writer Alphonse Allais, whose newspaper columns, written well over a century ago, were extremely popular. In order to distance himself from his subject he used his fantasy, a quality of which he was not in short supply (take his discovery of the frosted glass aquarium for shy fish, for example). During one particular pub crawl, Allais found himself peeing against a wall with some friends. Blissful with pleasure, he fastened up his trousers and sighed, 'Ah, my friends, if only I were rich. If I were rich I'd never stop pissing!'

Contributing to Allais's pleasure to a not-insignificant degree was companionship. Men learn this as boys engaging in pissing contests, the only sport at which I excelled. As a kid I toppled empty tin cans from more than a metre away with youthful exuberance. It's been a long time since I've been able to manage that, now that this mechanism, too, has slackened. Like all the other muscles, the bladder also decreases in strength with the advancement of years; a few long and hearty sessions in the loo are replaced by lots of feeble ones. The bladder muscles enclose a cavity in which the urine from the kidneys is collected until it's time to urinate. As the bladder fills, the pressure increases. Sensory fibres in the wall give the nervous system the first sign when the urine level reaches 0.25 litres. The autonomic nervous system opens the locks—if you're toilet-trained, that is. Toilet-trained people exercise a certain amount of control from the brain on the autonomic goings-on lower down, but when the pressure exceeds 0.1 atmospheres there's no stopping the tide. If you're asleep, you can only hope your bladder wakes you in time. During the day, the urge is mainly hastened by like-minded company. If someone next to you is urinating, you'll have to pee yourself, as if your willie had ears. Sometimes it takes a bit longer for the stream to get going—out of embarrassment, probably—but once it does, you splash along happily with the rest. But you don't really need another person. A leaking tap is enough to rouse you out of bed. It's irresistible. The autonomic nervous system won't let itself be bullied by the central authority. It has a lot in common with the Pavlov reflex. If a dog is used to being given food when it hears a bell ring, it'll start drooling whenever it hears the bell, even without food. If splashing goes with urination, the autonomic nervous system says to itself, then urinating goes with splashing. Consequently, you can help children with urination problems by playing a tape of splashing sounds. But give the child headphones or you'll end up having to pee yourself. In laboratories for urine research in which mice are used they have an even simpler method for getting the urine flowing: bare mouse feet on a cold floor. Cold feet make the bladder muscles contract, even in humans, so suddenly you have to go, even if your bladder is almost empty.

Not everyone is toilet-trained, not by a long shot. In the Netherlands alone, one million people cannot properly hold their urine. Half these people are elderly, with disorders of which incontinence is only a symptom. The other half are babies and toddlers whose autonomic nervous systems are not yet under control. But there's also a category between these two, people who wet themselves while having a good laugh, when the muscles seem to have a mind of their own. For theatre directors, cabaret performances are particularly notorious for the wet spots on the chair seats.

Town and parish councils are also strictly opposed to having their property pissed on. The acid in the urine makes ancient stones crumble. But peeing in public, especially in the company of your mates after a long night of loud carousing at the pub, can be a source of special pleasure. A toddler couldn't wish for a more glorious way to let off steam, although sometimes there's a heavy price to pay. Drunks who urinate in the canals of Amsterdam in order to spare the city's monumental buildings can fall in the water and drown, especially at night when the townspeople are off the streets. Police officers who dredge up the bodies have little difficulty determining the cause of death: the guy's fly is wide open. Serves them right, say many of Amsterdam's burghers. They shouldn't do such filthy things.

Is piss dirtier than shit? It's a question of taste. But in addition to liquid and solid there's a third state, at least as offensive, by which filth can escape: gas. It's much more difficult to seal the body off from gas. You may piss like a bull and shit like a dredging machine, but you get rid of most of your waste in the form of gas. You defecate once a day and urinate maybe ten times, but you exhale at least 17,500 times, your whole life long—and once you've breathed your last, you're a goner. When you think of breathing, you think mainly of the oxygen that comes in, but after every inhalation you also exhale to remove the used gases. Unlike faeces and urine, your gas is brought in and carried away through one and the same opening. It's a distasteful thought, certainly when you realise that part of the waste gas you've just eliminated will be unavoidably inhaled in your very next breath (unless you run very fast perhaps). But the reality is even worse: every breath you take has already been in someone else's mouth—in the mouth of your friend or that of your enemy, in the mouth of a dental assistant or in the dental catastrophe of dirty old men who chew tobacco. Milk is pasteurised by law, but who's doing anything about our air? If you really want fresh air in your bedroom, shut the window, don't open it.

If all is well, the amount of gas going out of your body is equal to the amount coming in. Otherwise you run the risk of bursting, or deflating, like a balloon. Yet not every breath is successful. At some point, in either Creation or during the course of evolution, a design fault occurred. The next time you get up in the morning, look into the mirror and say 'ahh' so you can see into your throat. That's where your windpipe intersects with your food pipe. A life-threatening situation. It was different with our forefathers the fish. The gases from the water went directly into the blood and out again by way of the gills, as they do in fish today, without the risk of taking a wrong turn. Our lungs developed as a protrusion of our intestines, which is why air and food pass each other in the mouth and have to travel through the same throat. Consequently, with every bite you take there's a danger of food ending up in your lungs, or—even worse—air in your stomach; just try getting rid of that during a formal dinner. What the throat most resembles is a railway crossing that lets the food pass through first, and then the air. The crossing is guarded by the epiglottis, which switches back and forth between hunger and shortness of breath like an insane crossing gate. It's hard for people who believe in Creation to accept the fact that their almighty God turned out such shoddy work. They see their Creator as a great watchmaker who fashioned all the gears in our body to mesh perfectly. I'm certainly willing to go along with that idea, except in the morning when I'm standing in front of the mirror and saying 'ahh'. Then I think: He may be a great watchmaker, but my throat could definitely use a great plumber.

You can hear it when air ends up in the intestinal tract. It's like a badly maintained central heating system. In both cases the sputtering is caused by air bubbles suspended in liquid. You can hear the grumbling right through your stomach, even though it's actually taking place further on, in your intestines. The remedy is obvious: bleed the radiator. But where's the closest valve, and where's the bleed key? In your stomach. The air here can escape the same way it came in: out of the mouth via the oesophagus and the throat to join the fresh air outside. This is called burping (officially it's *borborygmus*—the most wonderful onomatopoeia I know). Babies are good at it. When they suck on the nipple they accidentally take in a lot of air that has to be expelled from the stomach. But adults also swallow more air than they can handle while they're eating, especially if they talk too much at the same time, eat too fast, play with their false teeth, or if they're nervous. They burp a lot of the air back up, like a baby, but you can't deliberately burp it away. To make yourself burp, you first have to swallow air to get the process going, and usually more goes in than comes out.

Once the air has passed through the stomach there's no turning back. The next valve doesn't come along until the end of the tunnel, at the anus. If it hurries, the swallowed air will get there within ten minutes, so the air from your lunch will arrive at the same time as the bread from your breakfast. How do you get rid of it? While there are facilities for defecating and urinating that you can properly retreat to, there are no toilets or urinals for the evacuation of intestinal gas. A fart isn't very different from a turd with the shit scraped off, but that doesn't make it any less of a nuisance. All you can do is hold your fart until a suitable moment comes along. But the pressure can mount quickly. A human being has to get rid of 0.75 to 1.5 litres of intestinal gas a day on average—enough to blow up a party balloon halfway. That's too much gas for a healthy intestine, so about ten times a day you blow off another portion. Or more than ten, way more. One patient from Minnesota farted an average of thirty-four times a day, with a record of 141. 'He had very few friends,' reported the *New England Journal of Medicine*.

Dutch military humour from Uruzgan.

Even in antiquity, warnings were issued on the dangers of suppressing your farts. After one of his guests died as a result, retching from bottled-up shame, the Roman emperor Claudius came up with a plan to prohibit such suppression. But he ran out of time to execute his idea, thanks to a murderous wife. 'God only knows,' sighed Montaigne centuries later, 'how often our entrails have brought us to the very brink of an agonising death because we refused to release a single fart.' But things aren't usually as dire in actual practice. Gases, like fluids, can be absorbed by the intestinal wall and carried off with the blood. In the lungs they regain their freedom with a single breath. In this way a fart that was intended for the anus escapes through the mouth anyway. Fortunately, these don't smell very much.

Air that isn't burped or doesn't pass into the bloodstream will meet the light of day via the anus sooner or later. The resulting farts come in two varieties: A farts and B farts. A farts don't stink. They consist mainly of the nitrogen, oxygen and carbon dioxide from the swallowed air. The carbon dioxide from the air is supplemented by the gases that are produced in the extinguishing of gastric juices by the bicarbonate from the intestinal and pancreatic fluids. Bacteria also give off carbon dioxide. Hydrogen gas is released in the fermentation of starch, sugar and fibres, and in fairly large quantities: 12.5 litres is normal. About two-thirds of it takes the shortcut through the intestinal wall and the blood; the rest is converted into methane in half the human population. This half have the right bacteria to do the job, which manage to survive despite the lack of oxygen in the large intestine. They do the same thing outside our bodies in the oxygen-deficient ooze of the swamps, where their farts bubble up as marsh gas. The other half of the human population, those without methane bacteria, release a lot of hydrogen through the anus in an unchanged state. But methane or hydrogen, it doesn't make any difference as far as flammability is concerned. Both gases burn nicely. Methane was the source of countless mine explosions, and hydrogen fed the flames that tragically consumed the Hindenburg dirigible. So it's never a good idea to experiment with setting your own farts on fire. A bridegroom-to-be who tries it at his bachelor party can end up in hospital with blisters on his backside and scorched arse hair. And how many farmers' sons have blown their chances at inheriting the farm by igniting the fart of a family cow? Burnt to a crisp.

Although its name suggests otherwise, marsh gas—methane—is

just as odourless as nitrogen, hydrogen, oxygen and carbon dioxide. Before it can call itself a B fart, a methane fart has to be spiced. A pinch is enough. So there are powerful odorants involved, including hydrogen sulphide. This gas is made from a vestige of hydrogen in combination with the sulphur from the breakdown of old, worn-out cells from the intestinal wall, or from plants such as onions, cabbage and garlic that contain large quantities of sulphurous amino acids. The food industry provides extra sulphur by adding this sulphite to bread, beer, wine, fruit juice, crisps and chips. It makes your food last longer and your farts stink longer.

There are other excrement smells that hitch rides with your farts. Indole, skatole and the methyl sulphides all contribute to a bouquet so strong that it can be detected in a concentration of one in a hundred million particles. The persistence of the smell has to do with ventilation, but it also has to do with the specific density of your farts. Floor van den Hout, writing in the journal *Quest*, calculated that density in an average compound at 0.92 kg/m^3. That's somewhat lighter than ordinary air (1.29 kg/m^3). So with every fart you release your body doesn't get lighter but heavier. The fart rises gently to nose height like a more lasting memory of a fleeting event. *Kurz is der Furz, lang der Gestank*, short is the fart, long is the stench.

Many farts are accompanied by a loud signal that lets everyone around know who the perpetrator was. The system is reminiscent of those horns with the flashing lights on old-fashioned Dutch canal barges. In the absence of marine telephones, skippers used to communicate with each other by means of horns. By using a kind of Morse code, they let each other know what their ship was planning to do: port, starboard, reverse or moor. But in the tangle of all those ships you'd have to know which one was doing the tooting. This problem was solved by pairing each audio signal with a flashing light. Similarly, you know where the stench is coming from by the sound of the fart, even though this particular tooter would rather have kept it secret. You don't always have control over the volume of your farts. It all depends on how much gas has escaped. The sound is produced by the vibrating of your anus, which works just like the rubber mouthpiece on a deflating balloon. The pitch is determined by the diameter of the anal canal. A narrow opening produces a high tone, a wide opening produces a low tone. Sitting doesn't make the tone higher but louder, especially if you do it on the toilet with the bowl as your sound box. Haemorrhoids have no effect on the pitch at all.

Because the sound is merely a physical phenomenon you can easily imitate it. Ever since the thirties, special 'whoopie cushions', designed on the basis of a deflating balloon, have been sold in so-called party shops. The balloon itself is flat, with an eloquent wooden mouthpiece. A forerunner was the musical seat, which came out in 1926. It consisted of a drum and a bellows that, according to the Johnson Smith mail order catalogue, sounded 'as if you were sitting on a cat'. In the age of the computer, even the whoopie cushion has been electrified. Hundreds of thousands of 'fart machines' have been sold in shops and by post. They consist of a small speaker that you place under the chair of the person you intend to embarrass. With the help of a time exposure lever, the machine makes a clearly audible fart sound at the worst possible moment for the person sitting above it.

Judging from their etymologies, French farts are more powerful than English or German farts. In all three languages, farts are named onomatopoeically—a word that imitates the sound it describes, much like 'bang' or 'boom'. In French, a fart is a *pet*, the 'p' at the beginning clearly indicating a loud pooty toot. The English and German words 'fart' and *Furz* have an 'f' instead of a 'p', which is quite a bit softer and less rude, although it's the silent farts that are feared for their aromatic power. The Dutch, with the soft 'w' in *wind* or 'v' in *veest*, are among the softies. But it wasn't so long ago that our forefathers, true men of valour, produced a genuine *poep* with two p's, pronounced like the English 'poop' but referring to the gaseous and not the solid object. If the modern Dutch insist on emphasising the brute force of an exceptionally thunderous fart, they resort to the expressive gutterals that are part of their language and let loose a *scheet*, with its (for English speakers) barely pronounceable *sch*.

In polite company one just hopes that the wind one passes will be as gentle as possible, but sometimes a powerful blast of the horn is needed as a statement—an insult or a battle cry, if not a declaration of war. According to the Greek historian Herodotus, King Apries of Egypt released such a bellicose fart in 569 BC. When Apries gave orders to have the mutinous general Amasis apprehended, Amasis raised his buttocks from his saddle and farted, telling the messenger that he could take *that* back to the king. The king was so angry that he had the messenger's nose and ears chopped off, at which the people rose up in revolt, the king lost, and the general, as the new sovereign, led his people through forty-four fat years. Six centuries later, few of the lessons of history had been learned in Judea. Shortly after the death of King Herod in AD 44, according to the Jewish-Roman historian Flavius Josephus, it was a fart that triggered a slaughter among the Jews:

> When the multitude were come together to Jerusalem, to the feast of unleavened bread, and a Roman cohort stood over the cloisters of the temple, one of the soldiers pulled back his garment, and, cowering down after an indecent manner, turned his breech to the Jews, and spoke such words as you may expect on such a posture.

The crowd was incensed and a riot broke out. The Roman procurator sent in his troops, with disastrous consequences for the Jews. 'The Jews were in a very great consternation, and being beaten out of the temple, they ran into the city, and the violence with which they crowded to get out was so great, that they trod upon each other, and squeezed one another, till ten thousand of them were killed, insomuch that this feast became the cause of mourning to the whole nation.'

~

Yet this was but an incident compared with the disasters that flatulence has had the power to cause worldwide. One fart is irksome at the very most, while all our farts together are contributing to the greenhouse effect, and if you include the farts that animals produce it's clear that we're blowing the entire atmosphere to smithereens. By exhaling carbon dioxide, humans and animals make a substantial contribution to the air pollution that retains the earth's warmth like a layer of the atmosphere, thereby throwing the climate into a turmoil. But the methane in all those farts and burps is twenty times worse than the carbon dioxide that usually gets all the blame. The methane from humans alone is more than the atmosphere can deal with, but add to that the methane from cows and it's just too much. Cows produce a hundred times more methane than humans do. Dutch livestock alone fart and burp 100,000 tonnes of methane a year into the air. In the United States, cattle account for a quarter of the methane pollution, or 5 per cent of the greenhouse effect. In order to produce all that greenhouse gas, the cows in turn burn up 5 per cent of their feed, which can only be cultivated with the help of oil-guzzling, artificial manure-spreading and atmosphere-destroying agricultural machines. But the animals are capable of a lot even without our help. Three giraffes produce as much methane as one cow, and one elephant beats four cows to pieces. To say nothing of a whale. In his answer to a question from the audience about whether whale farts can be heard on his ship's sonar, marine biologist Richard Martin started working out the calculations. Endangered or not, all the whales together fart a gigantic 200 billion litres a day. This filled Martin with both emotion and awe. 'To know that whales fart, too, brings us closer together.'

Yet it isn't the giants but the dwarves from whom we have most to fear: the termites, better known from travellers' tales as white ants. These are the insects that come marching in enormous hordes to consume all your woodwork, preferably from the inside out, so that only the outside is left standing until the whole thing—*kaboom!*—comes crashing down. If you're still upstairs you'll soon be downstairs. The books of my childhood were quite clear on this subject. Men with wooden legs in particular were advised to watch their step in white ant country. Fortunately in the Netherlands it's too cold for white ants. Up until now, that is. With their contribution to the greenhouse effect, the termites are making it more comfortable for themselves, even in our corner of the world. They're helped by the microbes in their intestines, which digest the wood for them. For every bite they take they release a little fart. Because there are millions of them, all those mini-farts form a fart of impressive proportions, and with all the termite hills in the world the climate will ultimately be assaulted by a million times a million times a million farts. To be precise, the 2,500 billion termites on earth pump out 100 million tonnes of methane a year. 'That's a whole lot of methane,' says one expert, 'but it's also a whole lot of termites.'

The problem with farting is you so rarely can do anything about it. Before you really feel the urge it's often too late. Banning it doesn't help. Lawmakers, who are human themselves, know that all too well, but that doesn't stop them from issuing prohibitions. In Malawi a bill was introduced in 2011 authorising traditional leaders to punish inhabitants 'who foul the air'. In response, American actress Whoopi Goldberg announced that she wasn't going to travel to Malawi, at least not for the foreseeable future. She proposed an amendment to the bill: 'He who smelt it, dealt it'. Whoopi should know. She owes her first name to her childhood friends, who called her that because she farted so much. 'I was a walking whoopie cushion,' she wrote in her autobiography. 'I always call my farts tree monkeys, 'cause tree monkeys make the same farty sound as I do.'

How right she is. If you appreciate this kind of humour, your rectum and anus are always there to show you a good time. How many boring classes or lectures have been animated by a well-timed fart! There's nothing more likely to induce a glorious giggle. And it's an endless source of jokes. (As long as you like that sort of thing.)

> A guy goes to the doctor. 'Doctor, I can't stop passing gas.' The doctor considers the problem, walks to the back of the office and returns with a six-foot stick with a massive hook at the end. The guy blanches. 'What are you going to do with that thing?' 'Open the window,' the doctor answers. 'No time to lose. I'm practically suffocating!'

A fart doesn't amount to much compared with shitting and pissing, as a farmer in Goderich, Ontario, found out. Struck by an overpowering need to go while driving, and without a loo in sight, he pulled his car over, jumped out, and did his business. He was caught and fined eighteen dollars. Relieved by the small amount, the man handed the judge a twenty. 'Keep the change, your honour,' he said. 'Who knows, I may need to fart someday!'

A veterinary tour de force from 1743.

But sometimes it's just not funny anymore, and farting becomes a condition that has to be dealt with. A doctor can help, but the problems are usually more social than medical. Remedies are correspondingly more domestic.

To expel less air, you can try swallowing less of it. But that only eliminates the farts that were tame to begin with. We're talking about the gale-force winds, the real stinkers. How do you get them under control? How do you make the billions of bacteria in your intestines so respectable that they only produce benign gas, and then in dribs and drabs? By putting them on a diet. Limit their food supply. The more of your food you consume yourself, in your small intestine, the less there is left over for the bacteria in your large intestine to ferment. Unfortunately, this means putting yourself on a diet. To stop the big farts in their tracks, you have to avoid certain kinds of foods that you don't have the enzymes to deal with. One notorious example is milk. It contains a protein, lactose, that must be broken down in the small intestine by the enzyme lactase. Babies have lactase in abundance. Otherwise they wouldn't be able to live from their mother's milk. But some people experience a drastic reduction in the formation of lactase as they grow into adulthood. For Chinese, Arabs, Aborigines, Greeks and Italians, tolerating milk becomes increasingly problematic. It gives them stomach aches and diarrhoea, and makes them extremely sensitive. The fact that so many Europeans can tolerate milk probably has to do with their long experience with cattle breeding. Cattle breeders who could tolerate milk were one step ahead in the struggle for existence.

Unfortunately this never happened among tillers of the soil. Even after ten thousand years of horticulture, humans still don't have the enzymes at hand to digest vegetables properly. They can't deal with fibre or certain stubborn sugars, the oligosaccharides. These pass through the large intestine undigested, as food for the bacteria with their gas-rich eating habits. To silence a bubbling anus you'd have to give up vegetables like cauliflower and Brussels sprouts, onions and artichokes, radishes and lettuce, tomatoes and cucumbers. Wholewheat bread, soft drinks and sweeteners are also taboo. Even fruit can do harm: Adam's first sin after eating the apple may very well have been the first fart. In addition to fibre, an apple contains a lot of air.

Beans are the worst culprit. Nothing contains so many proteins or so much fibre as oligosaccharides. The oligosaccharides in beans consist primarily of verbascose, but the fart of all farts comes from the bean of all beans, the soy bean, which is specialised in raffinose and stachyose. Every variety, every strain of bean has its own fart capacity. Actually this should be noted on every can of beans in addition to the other nutritional information, complete with the delay factor, so you know how long you can expect to be regarded as pleasant company after a meal of beans. In the case of brown beans, it takes six hours for the storm to reach its peak. For his Western parody *Blazing Saddles*, director Mel Brooks had to take considerable artistic freedom to get his cowboys to erupt in a barrage of farts right after a traditional meal of brown beans, which put all ordinary shoot-outs to shame.

Fortunately, not all vegetables are equally pneumatic, and each person has their own sensitivities. Those who have problems with flatulence—as a quarter of all Europeans and Americans do—can experiment with what does or does not agree with them. As for the remaining vegetables, you can try to tame them. With boiling water. That'll teach them. Boiling increases the digestibility so there's less food for the bacteria to fart. But be careful with cabbage. The longer you cook it, the more harmful sulphurous compounds are created, which you can clearly smell before you eat your Brussels sprouts and afterwards. By carefully regulating the cooking time you can control the balance between the before stench and the after stench.

Cooking is an attempt to break down the harmful substances, but you can also remove them beforehand. Let peas and beans soak in water the night before you plan to cook them. When you throw out the water, you're also throwing out some of the oligosaccharides (and some of the vitamins too, actually). From 30 to 90 per cent of them are dissolved this way, depending on the type of bean. A more cunning approach is to get the bean to eat its own oligosaccharides. Let it sprout. In order to turn its bean into a bean sprout, the seedling partially eats itself up. Unfortunately, when a bean sprouts not only does it get rid of undesirable substances but the sprout itself produces additional oligosaccharides. This trick works better with dairy products. The bacteria that make cheese or yogurt from milk have already consumed the lactose before it can trouble you. But you have to give them plenty of time to get their eating done; from the point of view of flatulence prevention, cottage cheese is a rush job.

If subtraction doesn't help, try addition. Countless remedies have been used down through the ages as carminatives, or drugs that relieve flatulence, from pepper to peppermint, from lemon to lavender oil. Most of them come from Asia, which is also home to the pneumatic kitchen. Cloves, nutmeg, cinnamon, ginger and cardamom are used to combat the flatulence produced by spicy dishes full of oriental seasonings. Asafoetida, better known as devil's dung, is thought to be especially effective. It's the gum extracted from the *Ferula asafoetida* plant, which itself smells of sulphur.

Of course there are all sorts of pills and powders available in health food stores and on the internet. Here, too, the carminatives are exceeded in both diversity and ineffectiveness by the aphrodisiacs, which are said to revive passion. Activated charcoal, sold under the commercial name Norit, absorbs the gas in the large intestine before it can do any harm, but in its enthusiasm it also swallows an array of medicines. And it turns your excrement a sinister shade of black. Scientifically, the most interesting are the remedies containing the enzyme that enable our bacteria, but not ourselves, to go after the feared elements in our food. By adding lactase drops from a bottle, for example, you can make milk digestible for people who normally cannot tolerate it, and you can calm bean-based flatulence by treating your beans to a few drops of the enzyme that breaks down oligosaccharides (alpha-galactosidase). But as long as we can't teach the body itself to make those enzymes, these are just stopgap measures.

The expectation is that the wind produced by beans and grains is only going to pick up. Now that more and more people are eating wholewheat bread, vegetables and beans instead of white bread, chips and hamburgers, for their own health and the health of animals, their intestinal bacteria are going to be increasingly boisterous in their celebrations. Vegetarians in particular are walking aerosol cans. Those suffering from flatulence have to make a choice in the kitchen between health and offence. A sensible person is willing to tolerate a little bother for the sake of his health. A wise person will even see the humour in being such a nuisance. Flatulence is fun. You're never bored with a fart up your arse. Across the globe a cloud of sulphur and skatole is rising from the literature to spice up human affairs.

If our civilisation began with the Greeks, then the first literary farts originated with the Athenian playwright Aristophanes. In *Clouds* (423 BC), he takes aim at Socrates's atheism. If there's no Zeus to make the heavens thunder, the old man Strepsiades asks him, where does the thunder come from then?

> Socrates:
> Have you not understood me then? I tell you, that the Clouds, when full of rain, bump against one another, and that, being inordinately swollen out, they burst with a great noise.
>
> Strepsiades:
> How can you make me credit that?
>
> Socrates:
> Take yourself as an example. When you have heartily gorged on stew at the Panathenaea, you get throes of stomach-ache and then suddenly your belly resounds with prolonged rumbling.
>
> Strepsiades:
> Yes, yes, by Apollo I suffer, I get colic, then the stew sets to rumbling like thunder and finally bursts forth with a terrific noise.

At first, it's but a little gurgling *pappax*, *pappax*! then it increases, *papapappax*! and when I take my crap, why, it's thunder indeed, *papapappax*! *pappax*!! *papapappax*!!! just like the clouds.

Socrates:
Well then, reflect what a noise is produced by your belly, which is but small. Shall not the air, which is boundless, produce these mighty claps of thunder?

Strepsiades:
And this is why the names are so much alike: crap and clap.

It wasn't until centuries later that the *papapappax* resounded in England in Geoffrey Chaucer's *The Canterbury Tales* (1386–1400). While Nicholas, an Oxford student in 'The Miller's Tale', is secretly sleeping with the beautiful (but married) Alison, his lecherous rival Absalon pops up under the bedroom window and begs her for a kiss. She promises to give him one, just to get rid of him. Possibly instigated by Nicholas, Alison sticks her hairy naked buttocks out the window. Absalon gropes in the dark with his lips and becomes suspicious when he discovers there's no nose between the supposed cheeks. He swears vengeance and goes to fetch a coulter from a nearby blacksmith. Thus armed, he comes back and asks for another kiss. That gives Nicholas an idea.

Now Nicholas had risen for a piss,
And thought he could improve upon the jape
And make him kiss his arse ere he escape,
And opening the window with a jerk,
Stuck out his arse, a handsome piece of work,
Buttocks and all, as far as to the haunch.
Said Absalon, all set to make a launch,
'Speak, pretty bird, I know not where thou art!'
This Nicholas at once let fly a fart
As loud as if it were a thunder-clap.

> He was near blinded by the blast, poor chap,
> But his hot iron was ready; with a thump
> He smote him in the middle of the rump.

For the greatest fun with the greatest filthiness, turn your sights to Germany and surroundings—and to the most unexpected quarter. Not only was Martin Luther a big fan of bathroom humour, but Wolfgang Amadeus Mozart got off on it as well. It's a good thing the audiences in the grand concert halls didn't know anything about the naughty letters Mozart had written to his mother ('Bye-bye mama, Your loving son / Just let go another one') and especially to his beloved cousin Maria Anna Thekla:

> Ma très chère Cousine!
> I now wish you a good night, shit in your bed with all your might, sleep with peace on your mind, and try to kiss your own behind. Oh my arse burns like fire! what on earth is the meaning of this!—maybe muck wants to come out? yes, yes, muck, I know you, see you, taste you—and—what's this—is it possible? Ye Gods!—Oh ear of mine, are you deceiving me?—No, it's true—what a long and melancholic sound!

From Ireland, finally, comes the limerick. The kind of humour that this poetic form depends on is a perfect match for shit and piss jokes. The punchline of the poem (if it's any good), driven along by the strict metre, explodes like a thunderous fart. A classic example is the 'Farter from Sparta'.

> There was a young fellow from Sparta
> A really magnificent farter.
> On the strength of one bean,
> He'd fart 'God Save the Queen'
> Or Beethoven's 'Moonlight Sonata'.
>
> He was great in the 'Christmas Cantata'
> He could double-stop the 'Toccata'

> He could boom from his arse
> Bach's 'B-Minor Mass'
> And in counterpoint, 'La Traviata'.
>
> The selection was tough, I admit,
> But it did not dismay him a bit,
> 'Til, with arse thrown aloft,
> He suddenly coughed
> And collapsed in a shower of shit.

The 'young fellow from Sparta' really did exist. Except he didn't meet his end quite so miserably, and instead of farting 'God Save the Queen' he performed 'La Marseillaise'. Born Joseph Pujol, he scored many a triumph as *Le Pétomane*. Dressed in white gloves, he would raise the tails of his coat and imitate the sound of a toad, a nightingale, a random dog, and a dog with its tail stuck in the door—all with a slightly anal accent. To the cry of 'Gunners! Atten-tion! Ready—Aim—Fire!', cannon shots would ring out through the Moulin Rouge, the centre of Oh-la-la in *fin de siècle* Paris and recognisable by the red windmill out the front ('The blades of the Moulin Rouge—what a great fan for my act!'). The Pétomane even imitated farts. First was that of a little girl, then of a mother-in-law, followed by that of a bride on her wedding night (quiet) and on the morning after (considerably louder). To end his act, the Pétomane smoked a cigarette anally, played the anal flute and accompanied the singing of the enthusiastic audience.

Invariably they clapped for the wrong half of the act. The trick isn't blowing out a candle or blaring all the notes of 'Au clare de la lune' but replenishing the air supply. You'd never be able to fart your way through 'God Save the Queen' if you ran out of breath by the time you got to 'gracious'. But where to store an adequate supply of onions and beans? We happen to know how Pujol did it, thanks to a medical study from 1892. First, he blocked his breathing by bending himself almost double. Then he used his stomach wall to draw in outside air through his anus, as we do orally through our mouths with our chests. He was able to keep this up evening after evening. Encouraged by his success, he struck out on his own in 1894, for which he was sued for breach of contract. From *Le Petit Journal*: 'At first the director of the Moulin Rouge thought of going after him and bringing him back with a few well-placed kicks in his... musical domain; but on second thought, fearing the harm he might do to the instrument, he decided to take the case to court and let the judge stick *his* nose in it.' The director won, but Pujol's vengeance was sweet. He unmasked his replacement—*La Femme Pétomane*—as a fake fartiste with a pair of bellows under her skirts. The *Pétomane* farted on alone with mixed success until the First World War, when gas was no longer a laughing matter. Resuming his old trade, he became a baker again. It wasn't until the end of the Second World War that the 'Nightingale of the Moulin Rouge', age eighty-eight, breathed his last breath. A very quiet one. Nothing unusual about it.

10

Fun and Games

Shitting is a paradox. You make something that you actually want to get rid of. So the fun has to come from the shitting itself. And it does; all you have to do is enjoy it. Senses aplenty. Although there are no taste buds in your backside, you can tell the difference between one turd and another by the consistency, the creaminess and the shape, the way you can tell the difference in your mouth between custard and porridge, between hot chocolate with skin or without. Isn't it delightful to let a well-lubricated turd slurp through your half-relaxed anus like a cake of soap through your hand? Enjoy how the tension mounts, sometimes to the breaking point, before tapering off when the passage triumphantly gives way and the turd slides through. Suddenly you remember how deliciously spicy your meal was day before yesterday. All you have to do is sit down, give yourself plenty of time, and focus your attention like a wine connoisseur or a tea ceremony devotee. The more advanced practitioners go to the loo with the eager expectation that ordinary people reserve for entering a fine restaurant.

The only condition attached to minor scatological pleasure is that you be open to it. You have to be in the mood. *Aus einem traurigen Arsch*, as every German knows, *fährt nie ein fröhlicher Furz*. From a sad arse, never a cheerful fart. A good-humoured arse, on the other hand, is a wellspring of entertainment. Thanks to all the nerve endings located there, it's sensitive to stroking, kissing, licking or a good wallop. And as far as that's concerned it's the perfect counterpoint to the mouth, which also has other responsibilities but for that reason is not insensitive to the attentions of a tongue or a pair of moist lips. 'Lick my arse' or 'you can kiss my butt' is both a curse and a reference to the most intimate of all intimacies. In the Middle Ages, it was believed that heretics kissed the devil's backside at ungodly hours. Witches were cleverer. They worshipped the devil in the shape of a black cat; no livelier little bottom in existence—so kissable. According to reports, this was when the witches danced their witches' dances and sang their witches' songs. But it took Wolfgang Amadeus Mozart to compose really heavenly music to honour the anus, with lyrics he wrote himself such as '*Leck mich im Arsch*' (KV 231) and '*Beym Arsch ist's finster*' (KV 441b). In 1787 he disguised his appeal in Dog Latin: '*Difficile lectu mihi mars et jonicu jonicu difficile*'. This three-part canon (KV 559) was a good-natured pitfall for the tenor-baritone Johann Peyerl. When Peyerl sang it in his Bavarian accent, the German public clearly heard '*leck du mi im Arsch*', while Italians recognised the word '*cujoni*' (balls) in the repeated *jonicu jonicu*. This was immediately followed by the choir with '*O du eselhafter Peierl*' (KV 560a).

How a tongue can end up in an anus is anybody's guess. Normally it's a neighbouring organ that enjoys the occasional lick. The genitals are located right next to the piss and shit organs. 'But Love has pitched his mansion in / The place of excrement' wrote the Irish poet William Butler Yeats, in a distant echo of Saint Augustine's comment that 'we're born between faeces and urine'. Sewer and portico are so closely interwoven that biologists speak of one urogenital system, and if you have sperm problems you're sent to the hospital urologist. No wonder your tongue tends to stray there in the dark. But even in broad daylight, the similarity between an anus and a vagina is hard for anyone to miss. Gay men have little choice, and, if you can believe contemporary pornography, more and more straight men are making the occasional detour on the way to the main entrance. From the outside it's impossible to see that the anus is meant for anything other than an exit. The valve that closes off the entrance is located deep inside, near the three-way connection where the small intestine, the appendix and the large intestine meet. The penis seems to have been created for penetration of the rectum. It's shaped like a turd and has approximately the same dimensions. What comes out should be able to go back in. The only thing that's missing is lubrication. Since the Creator had another opening in mind—the vagina—that's where the grease nipples are located. Fortunately His creatures invented lubricating jelly.

A man is lucky. He has a penis and an anus, which means he can mount and be mounted. Apart from his mouth, his anus is the only possible access point for penile penetration. Most men leave this option untried. But if an anus can play vagina, a finger can easily take the place of a cock. In *Wetlands*, Charlotte Roche learned this at an early age from 'a really old lover':

> He wanted me to experience everything about male sexuality so that in the future no man could ever pull one over on me. Now I supposedly know a lot about male sexuality, but I don't know whether all of what I learned applies to all men or only to him. I still have to see. One of his cardinal rules was that you should always stick your finger up a guy's arse during sex. Makes him come harder. So far I can certainly concur. It's always a hit. They go wild. But you shouldn't discuss it with them beforehand or after.

Otherwise they'll worry they're gay and get all uptight. Just do it and afterward pretend nothing was ever in there.

Anyone who appreciates the fun of sex automatically discovers the pleasures that shit and pee have to offer. In the 69 position all your senses come in intimate contact with a landscape where sexual hillocks change imperceptibly into anal craters and urinal geysers, cloaked in lakes of mucous and indefinable smells that would have revolted you under other circumstances, but here and now exert a magical attraction. Being sexually active means getting smeared with shit, but so what? You're there anyway, right? If you're pinching the cat in the dark, you might as well give the dog a squeeze. A situation that stimulates the one will entice the other. Once you're in bed, the move from the genitals to their neighbours in the crotch is almost instinctive. Conversely, the switch from shit and pee to sex often takes place in the loo. All alone, in the safety of the toilet, you can satisfy two needs at once. Paper, drain pipes, privacy and wandering thoughts are all within reach.

Most people are prepared to put up with a bit of excrement or urine during sex with varying levels of pleasure, but there are some who see it as the sauce and whipped cream of sexual cuisine. In *Portnoy's Complaint* by Philip Roth, Alex Portnoy's girlfriend talks about a former lover who liked to watch her defecate on a glass-top coffee table while he lay on the floor beneath her. The newly married man in the film *Where's Poppa?* satisfied his desires just as shamelessly. After having sex on their wedding night, the young bride notices that her husband has done a 'caca' in bed. When she demands, 'How could you?!' he answers, 'Doesn't everyone?' James Joyce, the author of *Ulysses*, was luckier with his wife Nora. Their marriage, as his letters attest, was sealed with a mutual longing for shit:

> The smallest things give me a great cockstand...a little brown stain on the seat of your white drawers...a sudden immodest noise made by you behind and then a bad smell slowly curling up out of your

> backside. It must be a fearfully lecherous thing to see a girl with her clothes up frigging furiously at her cunt, to see her pretty white drawers pulled open behind and her bum sticking out and a fat brown thing stuck halfway out of her hole. You say you will shit your drawers, dear, and let me fuck you then. I would like to hear you shit them, dear, first and then fuck you.

Charlotte Roche also knows the power of a lick of shit. Some of her lovers 'like it when the tip of their cock has a little crap on it when they pull it out after butt fucking—the smell of the crap their cock's pulled out turns them on.' For such a man, 'there's crap ready to be found just a few centimetres inside the entrance. He only has to have stuck it in for a second and come in contact with the crap. Then when he pulls it back out and we try out another position, his cock functions like a fluttering crap-scented air freshener.' If the man she's with isn't fond of anal sex with the faeces included, Charlotte cleans her rectum three times beforehand 'until there are no more mini-chunks of crap visible. I'm perfectly prepared for clean butt sex, like a blow-up doll.' But you have to be careful. There's a known case of a man who was doing it with such a doll when fate stepped in: 'No sooner had I bit her in the neck than she farted and flew out the window.'

Sex shouldn't be clean. There should be something slightly fishy about it, literally as well as figuratively. Some men are perfectly happy with the smell alone. You see them in Amsterdam's Red Light District, standing in the queue at the reeking urinals. In the standard work *Die sexuelle Osphresiologie* (1906), Albert Hagen describes the case of an established notary who 'since his childhood had been known as an eccentric and misanthrope'. By his own admission, he 'stimulated his sexual urge' with the help of 'a number of sheets of toilet paper that he himself had used, which he then spread out on the blankets and looked at and smelled until he got an erection, *die er dann zur Onanie benutzte*'. After his death a large basketful of these papers was found next to his bed, each one carefully marked with the date and the year.

~

If it were purely a matter of humans and their lusts, all bodily orifices of sufficient dimensions would be used for sexual pleasure. The only ones to object were the gods. The church has always threatened hell and damnation to those who use their openings unproductively. It hasn't had much to say about faeces and urine as such, but the entanglement with sex makes improper use not only obscene but profane, which puts it squarely in the middle of the church's field of operations. Sex from behind has been a special target of the church's ire. Those who do it doggy fashion—*more canum*—were doomed by the mediaeval church fathers to burn in hell for eternity while riveted together in this position. But who of us today allows the church fathers to peek under the sheets? Modern individuals don't let the church tell them what to do. A modern individual doesn't believe in God anymore but in science. And in the scientific study of shit and sex, there's only one high priest who towers above all the others. Here the word of Freud is still the law.

Although even Sigmund Freud was unable to untangle sex from shitting, he did impose some kind of order. He saw sex the way a good Christian sees the Holy Trinity: one God in three persons—God the Father, God the Son and God the Holy Spirit. According to Freud, your sexual development begins at birth and follows three steps: the oral, the anal and the phallic phase. For the first year and a half, pleasure consists mainly of sucking on your mother's nipples, and after you turn three, sex comes to reside in the organs for which it is named. But in the intervening years, satisfaction gurgles in your backside. This anal phase is a real earthly paradise. Shitting to their heart's content, your intestines furnish you with continuous pleasure. For thousands of years, babies had to make do without rattles or teething rings, but their intestines never abandoned them. Even this paradise has its Fall, however. Suddenly the fun and games are over: you have to do your business on the potty. You may have been mama's sweetheart, but from now on you have to perform first before she presses you to her bosom. If you don't make it to the potty on time you'll never grow up. 'You want to be a baby your whole life?' And if you do make it, the grateful hug you get for your latest production is followed by the ultimate insult: your turd, the most beautiful thing you can make, a bit of yourself, the perfect means of self-expression, is hastily carried off and flushed away. Because it's dirty. Yuck. Naughty. Bad. If there's one place where a child loses its innocence, it's on the potty, according to Freud. Suddenly the world is divided into good and bad. Pride makes way for shame. That's what a Fall is all about. But a fall from what? What can a baby have done wrong?

It got older, that's what. It's like being guilty of adolescence when you're fourteen or deserving of the death penalty when you're a hundred—a child is condemned to being toilet-trained just because of its age. If the child is ready for it, the punishment isn't really so bad. You've got to come face to face with evil sooner or later. If enough time has been set aside for this encounter, Freud will let you move on to the next stage of life, the phallic, and grow up to be a fairly well-balanced individual. So the child is in no hurry; it isn't up to him. It's the parents who'd like him to hurry up and be toilet-trained. When I was born, in 1946, a child who wasn't toilet-trained was an enormous inconvenience. All my nappies had to be washed by hand, wrung out and dried. Even more intolerable was the idea that there was something wrong with a child who took longer to be toilet-trained. Children were expected to meet certain expectations. A dirty child was a discredit to its mother. To get that shit out where it belonged, children were locked up, threatened and scolded. Big suppositories were jammed into small bottoms. The result was usually counter-productive. The worst mistake was to start in with potties far too early. A baby simply doesn't have the nerve bundles to bring its sphincters under control for the first year and a half. At age two, only half of all babies are toilet-trained, and one year later 5 per cent are still letting their shit run free. Why does it take so long? Why do humans take three years to do something that cats only need three weeks to master? Kittens are using the litter box long before the baby of the house of the same age is out of nappies because that's more or less what they've done for millions of years. Their forefathers were burying their excrement long before the human, let alone the litter box, had been invented. Their behaviour is anchored in their genes. Our forefathers, by contrast, were living in the trees, where you couldn't bury anything and there was nothing to bury anyway, because you could just let your excrement drop without giving it a moment's thought. Our babies still do that, lying close to the ground in their cradles—until nature takes pity on them, pulls out all the stops, and improvises a nerve connection between arse and brain. But what about the other apes who have come down from the trees? A baby chimpanzee clings to its mother all day long, inundating her with piss and shit. But when it's old enough it gives her a warning sign. Without any prompting from its mother, it makes a little growl and sticks out its rear end, neatly directing its piss and shit elsewhere. It can't be a coincidence that the chimp starts doing this after the age of two, which is approximately when human babies become toilet-trained. Sensible human parents patiently wait until their child lets them know the time has come. Of course they get the potty ready, but they don't exert any pressure to bring about the inevitable. No pooing lessons are needed. Why should a child learn to do something it's been able to do perfectly well since its birth? Human children toilet-train themselves, too.

Chimpanzees have never heard of Sigmund Freud, but we have. Even those who haven't read his *Charakter und Analerotik* (1908) fear his frustrations and fixations. At any point from cradle to nursery to school, something can go wrong in the sequence of oral, anal and phallic phases. If the time for the next phase arrives while the previous phase has yet to be concluded, you'll be stuck. Should such a fixation take place during the oral phase, not only do you keep sucking your thumb and whining, but later on, as an adult, you take to drink, you eat too much and you adopt a dependent attitude for the rest of your life. If you're toilet-trained too early, you become anally fixated and develop an anal personality. According to Freud, that makes you tidy, stingy, and stubborn. Once it was just your shit that you tried to control; later on it's your whole life. It often begins with collecting. Football cards, shells, or little plastic dolls become cherished possessions that you can organise into a unified whole. There's nothing wrong with that. On the contrary, every science once began as a collection—of objects or of facts. But occasionally older people will revert to the primal practice of collecting excrement and find themselves in the colourful collection of *Abarten des Sexualverhaltens* as uro- or coprophiles. According to *Abarten* collector Günther Hunold, a urophile collects

> other people's urine in small, carefully labelled bottles. It is essential that the label contain not only the place and the date but also a personal description. At moments of extreme sexual arousal, a favourite bottle is opened and a drop of urine is spread on the penis or clitoris. This is followed by a lengthy session of masturbation.
>
> Coprophilia involves collecting the faeces of the partner. After being dried, the excrement is stowed away and labelled, as with the urophile.

While it rarely gets this extreme, the more common thing collected by an anal personality is hardly less vulgar: the dross of the earth. Instead of faeces, he accumulates money; instead of parental pride he reaps social prestige. Rather than give his money away he hoards it, meticulously counted, with place and date noted in a ledger. A real miser is totally wacko. He'll never understand the comment made by Sir Francis Bacon (1561–1626), that money is like manure, not in the Freudian sense but because it is 'of very little use except it be spread'. The miser isn't alone in his incomprehension. Our entire capitalist society is anally fixated—at least according to Norman Brown. It's no accident that his *Life Against Death* came out in 1959, at the height of the battle between capitalism and communism. The anal fixation argument allowed him to depict capitalists as even worse than leeches; now they could be dismissed as anal perverts. The only thing that helped, according to Brown, was to recognise the problem. To tame the beast in man you first have to accept it. As long as you don't admit that you're stuck in a depraved body, the door to a better world will remain closed. Every trace of shame must be expelled, beginning with shit.

Nonsense. Poppycock. That's what modern pseudo-intellectuals and head-shrinkers have to say about the anal personality. A strongbox is not a toilet. You can't blame your parents because you're a narrow-minded, fastidious know-it-all. Psychiatrists are perfectly aware that orderliness, stinginess and stubbornness often occur together, but they have another name for it—obsessive-compulsive personality disorder (OCPD)—and toilets or faeces are not part of the picture. In practice, a causal connection has never been demonstrated between bad toilet-training experiences and OCPD. The anal fixation is something Freud dreamt up.

Yet a connection between childhood and anal personality traits is undeniable. If interest in the anal is childish, it's a perfect fit for our age of infantilising. With their fondness for sports, games and amusement, people love to return to their childhood. Never before has the call of Jesus Christ to 'become as little children' struck such a sympathetic chord. Dressed in their shorts, adults merrily run after a ball, lick ice-cream cones, shop for the fun of it. If earthly existence was once the crucial test for a better hereafter, it's now become a game, a way to pass the time. And what better toy could you wish for than a butt full of shit: always available, absolutely free and with an opening like that on a party balloon?

If collecting is the basis of science, then play is the origin of art. And if the playful use of material begins with shit, then art is based on shit as well. No modelling clay models so well. And so gratefully. Shit is raw material and product rolled into one. You can show a good-looking turd to your mother as a readymade ('look, Ma, no hands!') or work it into a ball, a cake or a slide for dolls. You can stick little flags in it or birthday candles, you can mould a castle out of it, and finally you can watch it gurgle as it's being flushed away. But there are other possibilities. Once you catch the creative bug, nothing can stop you from composing a song, writing a book or putting together an atomic bomb with the same pleasure. Nothing resembles the creative process more than excrement: from the first urge, then on to moaning and straining, and then the triumphant finale—the difference being that you can show the final product off to everyone, even if you're not totally satisfied with it yourself. Thus Michel de Montaigne (1533–1592) described his own *Essais* in *De la vanité* as 'the excrement of an old spirit, first hard, then soft, and always indigestible'. Centuries later, Stephen King, in his autobiographical work *On Writing*, showed his gratitude for the anal powers of his youth, which were still responsible for the heat of his writing. He had once had a big fat woman as a babysitter.

> Eula-Beulah was prone to farts—the kind that are both loud and smelly. Sometimes when she was so afflicted, she would throw me on the couch, drop her wool-skirted butt on my face, and let loose. 'Pow!' she'd cry in high glee. It was like being buried in marsh gas fireworks. I remember the dark, the sense that I was suffocating, and I remember laughing. Because, while what was happening was sort of horrible, it was also sort of funny. In many ways, Eula-Beulah prepared me for literary criticism. After having a two-hundred-pound babysitter fart on your face and yell Pow!, *The Village Voice* holds few terrors.

Play, when it ripens into art, is of vital importance. For humans, a song, a book or a bomb are what a tail is to a peacock or antlers to a deer. Men in particular tend to flaunt their creations. Those with the finest art or the funniest tricks are able to outsmart their rivals. Bursting with pride, the best artists let themselves be photographed by the gutter press along with their conquest. But whether they'd ever be allowed to show off their most beautiful turd in the four-star suite of the Hollywood Hotel is rather doubtful.

Making necessity fun: the best way to lighten up your journey.

Let alone their chamber pots full of pee. While biologists are willing to grant a certain measure of recognition to urinating as a means of securing territory, psychologists and psychiatrists see little good (or mischief) in it. Even Freud, who assigned such an important role to shit in the development of the anal character, dismissed the pleasure of urinating in 1908 as nothing more than 'urethral eroticism' that would lead to 'burning ambition'. Later, in 1930, he came up with the idea that primitive man had learned to suppress this pleasure only by stifling the infantile urge to put the fire out by pissing on it. But the mouth of the bladder was never given a psychosexual phase similar to the anal phase. The oral and anal openings apparently gave Freud all the opportunity he needed to make stuff up. For him, men and women were as good as equal, urethrally speaking.

In public, however, preference is given to male urination: there are more pissoirs than ladies' toilets. But in one respect men are at just as disadvantaged as women when it comes to public facilities. If they have to defecate, a pissoir isn't going to do them any good either. When it comes to shitting, a man is just as vulnerable as a woman. Along the motorways of Europe there are few opportunities to calmly pull over and defecate, probably because it's considered too expensive. You need someone to keep the thing clean, and that costs money. This led the authorities of Paris in 1981 to develop the 'Sanisette', an automatic toilet that cleans itself from top to bottom after each paid use. You see them in other parts of the world as well today, but the advance is slow. I don't think money is the main problem—physical relief is never too expensive if the need is great enough. What's more objectionable is finding yourself in a sealed mechanism that you can only hope will re-open, preferably before it has included you in its high-pressure hot-water clean-up. Here the thigmophilia of the toilet quickly turns to the claustrophobia of the submarine.

All things considered, nothing can beat your own loo at home. Only there can you find the peace and quiet you need for the pleasure of relieving yourself. In addition, since men and women aren't divided into LADIES and GENTS, they both get an honest look at the little idiosyncrasies of each other's sex. Women, for instance, are more sensitive to smells, streaks and stains. In practice this means they will be the first ones to clean the loo, even if most of the mess is made by men. Many men even refuse to hang up a new roll of toilet paper when the old one is used up. So women do that, too, although they're rarely thanked for it. Men argue that women are doing it all wrong. Women put the roll in the holder so the end hangs down against the wall. Men insist that the right way is just the reverse, since then the paper hangs far away from the wall and you can easily reach it to tear off a sheet without scratching your hand on the plaster. Whether this difference between the sexes is universal, and what purpose it serves, is not clear. Further scientific study is called for.

~

It's the kind of study that's perfect for children. They make use of the same toilet, but they're less self-conscious about it. When children reach the question-asking stage, the toilet is mainly a source of inspiration. Can you feel how many are coming? And what colour will they be? Why do I never see my father pooing? Does what you've eaten make any difference? Why do some farts stink and others don't? To become a real scatologist, all a child needs are a couple of tips. I included them once in a children's book:

> Farts smell good, but they're hard to see. You can't grab hold of them. That's because they're made of gas. Even though you can't see the gas itself, you can still make your farts visible. You do that in the bathtub. They look like a row of bubbles. And you can hear them, too: blub, blub. Now that you can hear them and see them, you can also catch them. Just hold an upside-down jam jar over them. But first, take the jar and fill it under water, then turn it upside down and lift it halfway out of the bath. If you do that, a little water will stay in the jar. No air can get in to drive the water out. But your farts can. They bubble up into the jar and push the water back into the tub. This means you've caught your farts—they're in the jar above the water. To keep them there, put the lid on the jar while it's still under water. That way you can smell them later on.

Scientifically inclined children will stick a label on the jar indicating what they had eaten the previous day. If they do that every day they can compare them: what food produces the best farts? If you have a duplicate, you can exchange jars with a friend. This is what Robert Provine calls 'small science' in *Curious Behavior*. It's not small because it's trivial but because all you need is your own body to do it. And it makes you laugh. Some people can't get enough of it, not even after their childhood years have passed. According to his life partners Woelrat and Teigetje, Gerard Reve saw it as an endless source of fun:

When company comes, Gerard usually can't help but show them his fart-in-the-glass act. He'll be a little nervous because of the visitors, which makes him gassy. Then suddenly he'll hold his empty wine glass up to his backside and catch his fart with a loud bang, after which he covers the glass with his hand. Then he takes a whiff through a crack in his fingers and shouts with delight, I got it! Damn, I got it! Really! I got it! Smell it! A nice, well-rounded, fat fart! And when the visitor recoils in disgust, he smells it again and says, Oh, come on. Give it a whiff. Just for a minute. Really! I got it.

A dirty body is a joy forever. Perfect for playing Mushroom. This game requires two players. As one of them farts, the other one tosses flour over his partner's backside with both hands. If the timing is right a mushroom cloud will appear, just like a nuclear test. If more people play you can hold regional contests or even set up a society, like the Free Farters from *L'esclavage rompu* (1750). 'Every Free Farter must act, speak, and bear witness in the spirit of the society. Newly admitted members are exhorted to fart unashamedly in their own homes, in the street, and in company, as propaganda for the society.'

Things were done more scientifically at the Academy of Lagado, where Gulliver found himself in a remarkable laboratory during his travels:

> I went into another Chamber, but was ready to hasten back, being almost overcome with a horrible Stink. My Conductor pressed me forward, conjuring me in a Whisper to give no Offence, which would be highly resented; and therefore I durst not so much as stop my Nose. The Projector of this Cell was the most ancient Student of the Academy. His Face and Beard were of a pale Yellow; his Hands and Clothes dawbed over with Filth. When I was presented to him, he gave me a close Embrace, (a Compliment I could well have excused.) His Employment, from his first coming into the Academy, was an Operation to reduce human Excrement to its original Food, by separating the several Parts, removing

the Tincture which it receives from the Gall, making the Odour exhale, and scumming off the Saliva. He had a weekly Allowance from the Society, of a Vessel filled with human Ordure, about the Bigness of a *Bristol* Barrel.

It's a promising branch of science. But you can no more reconstruct a turd's earliest days than you can predict what the wares of the greengrocer or butcher are going to look like after passing through the intestine. Two artists can use exactly the same paint and produce very different paintings. No two days are alike, no two moods are alike. Can you tell from a turd in what state of mind it was made? On Laputia they thought so:

> Another Professor showed me a large Paper of Instructions for discovering Plots and Conspiracies against the Government. He advised great Statesmen to examine into the Dyet of all suspected Persons; their Times of eating; upon which Side they lay in Bed; with which Hand they wipe their Posteriors; to take a strict View of their Excrements, and from the Colour, the Odour, the Taste, the Consistence, the Crudeness, or Maturity of Digestion, form a Judgment of their Thoughts and Designs: Because Men are never so serious, thoughtful, and intent, as when they are at Stool; which he found by frequent Experiment: For in such Conjunctures, when he used merely as a Trial to consider which was the best Way of murdering the King, his Ordure would have a Tincture of Green; but quite different when he thought only of raising an Insurrection, or burning the Metropolis.

Among the readers who missed the irony in Jonathan Swift's *Gulliver's Travels* were intelligence services such as the CIA and Mossad. During the Cold War, they managed to lay their hands on the excrement of world leaders such as Leonid Brezhnev and Mikhail Gorbachev. To keep friendly excrement from unauthorised eyes, George W. Bush carried his own toilet with him for the depositing of intestinal state secrets while travelling in Europe.

Heinrich Böll elaborated on this theme in his *Group Portrait with Lady* (1971). In this novel, Sister Rahel keeps detailed records of the faeces of the young girls in her convent school, which she carefully examines.

> The girls were required not to flush away these products into the invisible regions before Rahel had inspected them. In most cases a glance was enough for Rahel, enabling her to state with accuracy the physical and mental condition of the girl in question, and, since she could predict even scholastic achievement on the basis of excrements, she used to be positively beleaguered before the writing of term papers. Taking two hundred and forty school days as an annual average, times twelve girls and five years of floor-service, it is no trick to calculate that Sister Rahel kept statistical records and condensed analyses of some twenty-eight thousand eight hundred digestive processes: an astounding compendium that would probably fetch any price as a scatological and urological document. Presumably it has been destroyed as trash!

Since even graphology as a science is considered suspect, the reading of turds is not high on the list of academic methodologies. Of course, our turds want to tell us something—shitting is clearly a way of expressing yourself—but their message falls on deaf ears. Actually, it's more a mumble. Even the most modern bugging devices are unable to elicit a sensible syllable from a turd. You might as well resort to the oldest method known to natural science: tasting. You did it as a baby. Adults happily tuck into a nun's fart, a deep-fried pastry that's also popular in Germany (*Nonaffairs*) and France (*pet-de-none*). The name refers to a nun who was doing some cooking in the abbey of Armouries (near Strasbourg) when she farted. Startled with shock and embarrassment, she dropped a spoonful of dough into the cooking oil. Everyone was enchanted with the delicate result, which evoked vague memories of their earliest childhood. While proper ladies giggle as they take one last nun's fart with their tea, real men give a brazen wink and shove another sausage or meatball down their gullet. Or, big tough guys that they are, they eat some weird cheese that stinks like shit for miles around. But they wouldn't eat real shit for all the tea in China. Or so they think. Great is the indignation every time traces of faeces are found on the meat in a slaughterhouse. Yet it can't be avoided. When the meat was still a cow or pig, it was wrapped around a packet of intestines that were crammed full of shit. All it takes to puncture an intestine is for a knife to slip while the animal is being slaughtered. If this escapes the notice of the inspectors there'll be shit on your bologna sandwich. It is virtually impossible to maintain a strict separation between excrement and food. The entire food industry is simply full of shit. Vegetables and grains spend their whole lives under a layer of manure; cows walk around in their own food, which itself is covered in shit, and they never wash their hooves. You can rinse your vegetables off, of course, but the efficacy of rinsing is evident soon enough when you grate your teeth on the last bit of sand in your first bite of spinach. And even if it is possible to remove the faeces from a cow in plenty of time, intestines and all, how do you deal with a bag full of shrimp or a tin of sardines? Many small animals are eaten complete with bowel contents for the sake of convenience. It's true that boiling kills bacteria, but it's cold comfort to realise that the shit you've eaten was boiled, too. And sometimes you can't even count on the comfort—oysters are eaten alive, with living shit in living intestines.

'I swear it smells like violets!'
Political cartoon featuring Napoleon and his ever-faithful Marshall Michel Ney.

Everything has its fans. Hunters regard the excrement of snipes as a delicacy. When they clean them they deliberately leave the intestines intact in order to enjoy a bit of 'snipe dung'. It makes them feel more profoundly connected to nature. This kind of decadence goes back to the ancient Romans, of course. According to Lepidus, their emperor Commodus often ate excrement at his bacchanals. Hungarians would never do such a thing. They use their shit to brew brandy. At least that's what the Croats say. In Austria, the same claim is made about the Poles. At least according to coprologist Werner Pieper, a German.

Urine and shit may not be everyone's cup of tea, but there's a lot to be said for it nutritionwise. Your shit contains one-tenth of the calories from the food you eat, and your intestinal bacteria add their own valuable nutrients. There are pathogens in shit, too, of course, but not necessarily enough to kill you immediately. If eating your own shit made you drop dead, there would be considerably fewer dogs and pigs in the world. Yet most human shit-eaters don't do it for the nuritional value. A real coprophage is concerned with what's going on in his head. A morsel of excrement or a glass of piss can really turn him on. In the archive of Günther Hunold there's the case of Babsy Z.:

> Toni was a very nice man, although he had his preferences. But I can take a lot. As long as you pay I'll do anything for you.
>
> Except for hitting. I don't like that. But Toni wanted something very different. He made me drink a lot of champagne. I wasn't allowed to go to the toilet. 'Just hold it,' he said. Then he gave me more to drink. At a certain point I couldn't hold it anymore. I crossed my legs and pressed my thighs together. Then I noticed that he was starting to get aroused. He took this odd sort of bottle out of his black briefcase.
>
> Toni asked me to urinate into the bottle under the table. He wanted to make sure he could see everything, he said. While the bottle was filling up, he began to masturbate. Then Toni grabbed the full bottle. He took a golden dish out of his briefcase and poured the urine from the bottle onto the dish. Then he took out a

golden spoon and began consuming the liquid. At the same time he grabbed my hand and pulled it over to his erect penis. I knew what I was supposed to do.

Coprophagia reached its high point with the Marquis de Sade. In *Les cent-vint journées de Sodome* (1785) he writes about an orgy that lasted four months. It's a sampling of sexual tastes, from delicate to coarse. When it comes to excrement, each of the four guests has his preference. One of them, a young financier, only eats shit that's more than a week old, preferably with a bit of mould on top. A second is wild about diarrhoea, especially that of women who are suffering from indigestion or have just taken a laxative. There's a freethinker who goes to Communion and gets a pair of whores to shit on his host while it's still in his mouth. And then the fourth, a landowner, lives with a woman whom he keeps on a strict diet (no fish, salted meats, eggs, dairy products or bread, and very little fat, but plenty of poultry) to improve the taste of her faeces, which he dines on daily.

Emperor Commodus enjoying his coprophagic meal.

This puts him well ahead of his time. I see a future in which the shitting that will follow a meal will be taken into account as a matter of course, so you can enjoy the same meal twice. This doesn't mean actually eating your shit. There are so many other ways to have anal fun with your food. A day or two after the meal you can expect the only real dessert that counts, either gassy or with the consistency of conventional desserts like pudding, according to your taste. The true food connoisseur will also be a faecal connoisseur. Right now, you can already turn to Indian cuisine for the coarser work. Warnings like 'very, very spicy' or 'extremely hot' are what to look for. It's fire-eating followed by shitting razor blades, all within a space of twenty-four hours. This is no longer eating; it's big-time athletics. The strength of the peppers determines the height of the bar. For good Spanish peppers you don't even need taste buds; you can taste them with your skin. Try rubbing the inside of your wrist with a sliced hot pepper. It burns. You can imagine what happened to a friend of mine when she walked away from a pan of *ajam setam* for a minute, her hands still damp with pepper juice, to change a tampon. For the anus, it's a stroke of luck when something finally comes along that can be tasted without taste buds. At last, a chance to get involved. But it doesn't always have to be spectacular. Even for minor anal pleasure it's easy to put a menu together with the help of peas and beans, bananas, raisins, coconut milk, mangos, coffee and cigars. It's all just a matter of thinking ahead.

Tasting with your anus, eating out of your own butt—it sounds so unnatural. No wonder the church was against coprophagia. In the Bible, when Jerusalem was being threatened by the king of Assyria, the biggest insult to 'the men which sit on the wall' was 'that they may eat their own dung, and drink their own piss'. This was too much for the God of Vengeance. 'And it came to pass that night, that the angel of the LORD went out, and smote in the camp of the Assyrians an hundred fourscore and five thousand'. Another time, God was sorry when he had the prophet Ezekiel eat something that had come in contact with faeces:

> Take thou also unto thee wheat, and barley, and beans, and lentiles, and millet, and fitches, and put them in one vessel, and make thee

bread thereof, three hundred and ninety days shalt thou eat thereof. And thou shalt eat it as barley cakes, and thou shalt bake it with dung that cometh out of man, in their sight.

Then said I, Ah Lord GOD! behold, my soul hath not been polluted: for from my youth up even till now have I not eaten of that which dieth of itself, or is torn in pieces; neither came there abominable flesh into my mouth. Then he said unto me, Lo, I have given thee cow's dung for man's dung, and thou shalt prepare thy bread therewith.

The taboo on eating faeces continues unabated and has even intensified, owing to today's consumer society. When consuming is regarded as the greatest good, there's little room for a commodity as unsuited for consumption as your own shit.

Yet we owe our excrement more respect than we normally show. What we toss out is not worthless rubbish but a valuable product of intricate, highly synchronised processes. Not only is shit the keystone of the circle of life, but it's also an object of lust, a means of creative expression, and a vehicle for communicating love between humans or between man and animal. Shit is the universal lubricant of the entire life mechanism; shit keeps the whole thing going. One of the first to recognise this was the philosopher Diogenes (c. 412–323 BCE). The list of things he had respect for was very short indeed. He regarded nice clothing, good manners and grand banquets as futile attempts to distinguish ourselves from the animals. Shit, on the other hand, was something he could pay deference to. To underscore this, he hiked up his robe in the middle of the marked square and defecated the natural way. Respectable people have never been able to understand this, any more than they understood Gustave Courbet, famous for his painting *L'origine du monde* with its glimpse inside a beautiful nude woman as seen from below. When asked what he thought of a neatly painted little landscape by François-Louis Français, Courbet expressed his disapproval with the words 'There's nowhere to shit in it!' Don't go where you can't shit. Heinrich Heine once told an anecdote about the same high regard for excrement. While waiting for an appointment with a member of the Rothschild family, he saw a servant walk past who was carrying an elaborately decorated silver chamber pot, obviously belonging to his master. A young man who happened to be waiting with Heine tipped his hat in respect. That young man will go far, Heine said to himself.

So it would seem fitting to welcome your own faeces into the world with a certain amount of ceremony. Fortunately nearly everyone has the requisite setting for the family loo: a porcelain baptismal font, gleaming white and ready for the new citizen of the world. It may even give rise to an aura of serenity, most successfully achieved by the classic Japanese toilet and lyrically described by Jun'ichiro Tanizaki (1886–1965) in *In Praise of Shadows* as 'a place of spiritual repose'. 'No words can describe that sensation as one sits in the dim light, basking in the faint glow reflected from the *shoji*, lost in meditation or gazing out at the garden. Here, I suspect, is where haiku poets over the ages have come by a great many of their ideas.' But you don't have to be a haiku poet to unwind on the loo, staring into the distance and plunging into your own fantasies or those in a book. A toilet is a reader's throne. 'I've done all my best reading on the toilet,' Henry Miller confessed. For Marcel Proust, the little room 'intended for a more specific and vulgar use' was a place for 'all my activities that demanded inviolable solitude: reading, daydreaming, weeping, and sensual pleasure'.

Whenever I went to the library as a boy to borrow new books, I would look into the reading room with astonishment. Why would you sit in a room with other people and read when you could do it at home alone in the loo? Didn't these people have indoor plumbing? In a room like that your body has nothing to do; on a toilet your mind and body are concentrating together. It's not for nothing that the faces of people who are pooing look so much like those of people who are thinking. Squeezing out a brilliant idea requires the same effort as squeezing out a hard turd; you see the veins in the forehead swelling in order to transport extra blood. Actually, the pose of Rodin's 'The Thinker' is no different from that of an equally concentrated Pooer.

Even a brief visit demands some decorum, all the more because there was so little time to get acquainted. Your faeces vanish from your life almost as soon as you give birth to them, on their way to an uncertain future. This can be a traumatic experience, especially for children. The six-year-old daughter of a Chicago psychiatrist became completely overwrought because each time a little piece of herself disappeared into the dark hole of the toilet. She could only be comforted by her father reassuring her that the little bit of human being went to be united with its ancestors in some cesspool paradise where everyone's excrement would live happily ever after.

Nothing for it but to say goodbye, something even grown-ups aren't very good at. When you've had a good visit with someone you look forward to leaving because you can always come back again, but the more unenjoyable the visit is the more difficult that becomes. Many couples stay together simply because neither one of them knows how to let go. The advantage of being a turd is that you barely have time to become attached before the moment of separation arrives. But when it does, we're forced once again to face the unbearable fact of life's brevity. Man is but a fart in the endlessness of eternity, but what a stinker he is! All that's left of us after our departure is a puff of smoke—all the more reason to pause and give some thought to the passing of a creature even more perishable than ourselves: our very own turd child.

'See you later,' you murmur, but you don't really mean it. 'Take care now,' and you pull the chain.

Bibliography

Ackerman, Diane, *A Natural History of the Senses*, New York: Random House, 1990.

Akkermans, L.M.A., *De hersenen in de darm: Vervolg*, Utrecht: Utrecht University, 1995.

Akkermans, Louis, *Gut in Motion*, Utrecht: Gastro-Intestinal Motility Center, AZU, 1996.

Alexander, R. McNeill, *Energy for Animal Life*, New York: Oxford University Press, 1999.

Allen, Valerie, *On Farting: Language and Laughter in the Middle Ages*, New York/Basingstoke, Palgrave Macmillan, 2007.

Altman, Dagmar, *Harnen und Koten bei Säugetieren: Ein Beitrag zur vergleichenden Verhal- tungsforschung*, Wittenberg Lutherstadt: A. Ziemsen, 1969.

Amato, Joseph A., *Dust: A History of the Small and the Invisible*, Berkeley/Los Angeles/London: University of California Press, 2000.

Aristophanes, *Peace, The Frogs*.

Aristophanes, *The Clouds*, http://classics.mit.edu/Aristophanes/clouds.html

Ashton, T.J., *On the Diseases, Injuries and Malformations of the Rectum and Anus, With Remarks on Habitual Constipation*, Philadelphia: Henry C. Lea, 1865.

Bang, Preven, *Elseviers diersporengids: Sporen en kentekens van zoogdieren en vogels*, Amsterdam/Brussels: Elsevier, 1973.

Barbara, Anna and Anthony Perliss, *Invisible Architecture: Experiencing Places Through the Sense of Smell*, Milan: Skira, 2006.

Barnes, David S., *The Great Stink of Paris and the Nineteenth-Century Struggle Against Filth and Germs*, Baltimore, MD: Johns Hopkins University Press, 2006.

Bart, Benjamin, *The History of Farting*, London: Michael O'Mara, 1995.

Bartels, Toon, *Kleine obstipatologie*, Amsterdam: Philips-Duphar Nederland, 1978.

Beaumont, William, *Experiments and Observations on the Gastric Juice and the Physiology of Digestion*, Platsburg, NY: F.P. Allen, 1833.

Benninga, M.A., et al. (eds.), *Het obstipatie formularium: Een praktische leidraad*, Houten: Bohn Stafleu van Loghum, 2010.

Blaxter, Kenneth, *Energy Metabolism in Animals and Man*, Cambridge, MA: Cambridge University Press, 1989.

Boersma, J. S. et al., *Latrines*, Meppel: Edu'Actief, 1994.

Bolin, Terry and Rosemary Stanton, *Wind Breaks: Coming to Terms with Flatulence*, McMahons Point, Australia: Margaret Gee Publishing, 1993.

Böll, Heinrich, *Group Portrait with Lady*, translated by Leila Vennewitz, New York: McGraw-Hill, 1973.

Bologne, Jean-Claude, *Histoire de la pudeur*, Paris: Olivier Orban, 1986.

Bomans, Godfried, *Pieter Bas*, Utrecht/Antwerp: Het Spectrum, 1973.

Boon, T.A. (ed.), *Basisboek urologie*, Leusden: De Tijdstroom, 2001.

Bourke, John Gregory, *Der Unrat in Sitte, Brauch, Glauben und Gewohnheitsrecht der Völker: Verdeutscht und neubearbeitet von Friedrich S. Krauss und H. Ihm*, Leipzig: Ethnologischer Verlag, 1913.

Bromcie, Alec, *The Ultimate Book of Farting*, London: Michael O'Mara, 2000.

Bruggen, Harry van, *Patiënt, privaat en privacy: De stoelgang als gezondheidswetenschappelijk probleem*, Lochem: De Tijdstroom, 1991.

Buch, Walter, *Der Regenwurm im Garten*, Stuttgart: Eugen Ulmer, 1984.

Buckman, Robert, *Human Wildlife: The Life That Lives On Us*, Toronto: Key Porter Books, 2002.

Burgess, Anthony, *Naar bed, naar bed*, Baarn: Bosch & Keuning, 1982.

Burke, David, *Bleep! A Guide to Popular American Obscenities*, Los Angeles/San Francisco: Optima Books, 1993.

Calvino, Italo, *Under the Jaguar Sun*, translated by William Weaver, New York: Harcourt Brace Jovanovich, 1988.

Campkin, Ben and Rosie Cox (eds.), *Dirt: New Geographies of Cleanliness and Contamination*, London/New York: I.B. Tauris, 2007.

Catalani, Carla, *Het bed: 5000 jaar waken en slapen*, Bussum: C.A.J. van Dishoeck, 1968.

Chappell, George S., *Through the Alimentary Canal with Gun and Camera: A Fascinating Trip to the Interior*, New York: Frederick A. Stokes Company, 1930.

Chaucer, Geoffrey, *The Canterbury Tales*, rendered into modern English by Nevill Coghill, London: Penguin Classics, 1951.

Chivers, D.J. and P. Langer (eds.), *The Digestive System in Mammals: Food, Form and Function*, Cambridge/New York/Melbourne: Cambridge University Press, 1994.

Classen, Constance, David Horves and Anthony Synnott, *Aroma: The Cultural History of Smell*, London/New York: Routledge, 1994.

Comfort, Alex, 'Communication May Be Odorous', *New Scientist and Science Journal*, 25 February 1971.

Corbain, Alain, *Pestdamp en bloesemgeur: Een geschiedenis van de reuk*, Nijmegen: SUN, 1986.

Creek, Wilson, *Now Here's a Man Who Knows His Shit*, Ontario, OR: Possum Press, 1999.

Crelin, Edmund, S., *Functional Anatomy of the Newborn*, New Haven/London: Yale University Press, 1973.

Dalgleish, Tim and Mick J. Power (eds.), *Handbook of Cognition and Emotion*, Chichester, etc.: John Wiley & Sons, 1999.

Dam, Johannes van, *Eet op!* Amsterdam: Nijgh en Van Ditmar, 2002.

Darwin, Charles, *The Descent of Man, and Selection in Relation to Sex*, London: John Murray, 1871.

Darwin, Charles, *The Expression of the Emotions in Man and Animals*, edited by Paul Ekman, New York: Harper Collins, 1999.

Darwin, Charles, *The Formation of Vegetable Mould, Through the Action of Worms, With Observations on their Habits*, London: John Murray, 1881.

Dawson, Jim, *Who Cut the Cheese? A Cultural History of the Fart*, Berkeley, CA: Ten Speed Press, 1999.

Dawson, Jim, *Blame It on the Dog: A Modern History of the Fart*, Berkeley, CA/Toronto: Ten Speed Press, 2006.

De balans: Functionele aspecten van de darmflora, Almere: Yakult Europe, 2001.

Dekkers, Midas, *Het edelgedierte: Over het vreemde verbond tussen mens en dier*, Amsterdam: Bert Bakker, 1978.

Dekkers, Midas, *Je binnenste buiten*, Amsterdam: Meulenhoff Informatief, 1986.

Dekkers, Midas, *Eten is weten*, Amsterdam: Meulenhoff Jeugd, 1988.

Dekkers, Midas, *Lief dier: Over bestialiteit*, Amsterdam/Antwerp: Contact, 1992.

Dekkers, Midas, *De vergankelijkheid*, Amsterdam/Antwerp: Contact, 1997.

Dekkers, Midas, *De larf: Over kinderen en metamorfose*, Amsterdam/Antwerp: Contact, 2002.

Delvoye, Wim, *Cloaca*, New York: New Museum of Contemporary Art/Ghent: / Rectapublishers, 2001.

Delvoye, Wim, *Fabrica*, Prato: Centro per l'Arte Contemporanea Luigi Pecci/Ghent: Rectapublishers, 2003.

Diepenbeek, Annemarie van, *Veldgids diersporen: Sporen van gewervelde landdieren*, Zeist: KNNV, 1999.

Dieren, Wouter van, *Een grondige zaak: 50 jaar Vuilafvoer Maatschappij VAM, 1929-1979*, Amsterdam: VAM, 1979.

Döpp, Hans-Jürgen, *In Praise of the Backside*, New York: Parkstone Press International, 2010.

Dorrestijn, Hans, *Het complete anti-hondenboek*, Bert Bakker, Amsterdam, 1992.

Doty, Richard L., *The Great Pheromone Myth*, Baltimore: Johns Hopkins University Press, 2010.

Douglas, Mary, *Purity and Danger: An Analysis of the Concepts of Pollution and Taboo*, London and New York: Routledge, 1966.

Drewry, George Overend, *Common-Sense Management of the Stomach*, London: Henry S. King & Co., 1875.

Drisdelle, Rosemary, *Parasites: Tales of Humanity's Most Unwelcome Guests*, Berkeley/Los Angeles/London: University of California Press, 2010.

Dundes Alan, *Sie mich auch! Das Hinter-Gründige in der Deutschen Psyche*, Weinheim/Basel: Beltz Verlag, 1985.

Duparc, Frederik and Quentin Buvelot, *Philips Wouwerman 1619-1668*, Zwolle: Waanders, 2009.

Ebberfeld, Ingelore, *Botenstoffe der Liebe: Über das innige Verhältnis von Geruch und Sexualität*, Frankfurt/New York: Campus Verlag, 1998.

Ehrenpreis, Eli D., Shmuel Avital and Marc Singer, *Anal and Rectal Diseases: A Concise Manual*, New York, etc.: Springer, 2012.

Ekman, Paul, *Gegrepen door emoties: Wat gezichten zeggen*, Amsterdam: Uitgeverij Nieuwezijds, 2003.

Elias, Norbert, *Het civilisatieproces: Sociogenetische en psychogenetische onderzoekingen*, Utrecht/Antwerp: Het Spectrum, 1982.

Emsley, John, *The Shocking History of Phosphorus: A Biography of the Devil's Element*, London: Macmillan, 2000.

Englisch, Paul, *Das skatologische Element in Literatur, Kunst und Volksleben*, Stuttgart: Julius Püttmann, 1928.

Everdingen, J.J.E. van (ed.), *Beesten van mensen: Microben en macroben als intieme vijanden*, Overveen: Belvédère, 1992.

Faber, René, *Von Donnerbalken, Nachtvasen und Kunstfurzern: Eine vergnügliche Kulturgeschichte*, Frankfurt am Main: Eichborn, 1994.

Fabre, J.H., *Het leven der insecten*, Amsterdam: Scheltema & Holkema, s.a.

Feixas, Jean, *Pipi caca popo: Histoire anecdotique de la scatologie*, Geneva: Editions Liber, 1996.

Feldhaus, Franz Maria, *Ka-Pi-Fu und andere verschämte Dinge: Ein fröhlich Buch für stille Orte mit Bildern*, Berlin: privately published, 1921.

Fendel, Sofie, *Dir gehört der Arsch versohlt: Die erotische Freude am Popoklatschen*, Cologne: Cologne, 1979.

Flanagan, Deuce, *Everybody Poops 410 Pounds a Year: An Illustrated Bathroom Companion for Grown-Ups*, Berkeley, CA: Ulysses Press, 2010.

Flaws, Bob, *Scatology & the Gate of Life: The Role of the Large Intestine in Immunity, an Integrated Chinese-Western Approach*, Boulder, CO: Blue Poppy Press, 1990.

Fox, Charles James, *An Essay Upon Wind: With Curious Anecdotes of Eminent Pêteurs*, Richmond, UK: Oneworld Classics, 2007.

Franklin, Benjamin, *A Letter by Dr. Franklin to the Royal Academy of Brussels*, New York: At the Sign of the Blue-Behinded Ape, 1929.

Franklin, Benjamin, *Fart Proudly: Writings of Benjamin Franklin You Never Read in School*, Columbus, OH/Atlanta, GA: Frog, Berkeley and Enthea Press, 1990.

Freud, Sigmund, *Civilization and Its Discontents*, translated from the German by James Strachey, New York: W.W. Norton & Company, Inc., 1961. Originally published as *Das Unbehagen in der Kultur*, Vienna: Internationaler Psychoanalytischer Verlag, 1931.

Freud, Sigmund, *Drei Abhandlungen zur Sexualtheorie*, Leipzig/Vienna: Franz Deuticke, 1905.

Furze, Peter, *Tailwinds: The Lore and Language of Fizzles, Farts and Toots*, London: Michael O'Mara, 1998.

George, Rose, *The Big Necessity: The Unmentionable World of Human Waste and Why It Matters*, Metropolitan Books, New York, 2008.

Gershon, Michael D., *The Second Brain: The Scientific Basis of Gut Instinct*, New York: HarperCollins Publishers, 1998.

Glas, G.A., *Volle boel: Het straatmeubilair van PTT temidden van al het andere*, The

Hague: Stichting Het Nederlandse PTT Museum, 1991.

Glaser, Gabrielle, *The Nose: A Profile of Sex, Beauty, and Survival*, New York: Washington Square Press, 2002.

Goldberg, Whoopie, *Book*, New York: R. Weisbach Books, 1997.

Goldschmidt, Richard, *Einführung in die Wissen- schaft vom Leben oder Ascaris*, Berlin/Göttingen/Heidelberg: Springer-Verlag, 1927.

González-Crussi, F., *The Five Senses*, New York: Harcourt Brace Jovanovich, 1989.

González-Crussi, F., *Carrying the Heart: Exploring the Worlds Within Us*, New York: Kaplan Publishing, 2009.

Gotthardt, Elmar, *Historia naturalis vaporum: Als Naturgeschichte der Fürze zu Deutsch verdollmet- schet, mit gelehrten Annotationibus illustriert, wohl gesetzet und ans Licht gestellt*, Urnäsch: Hanspeter Gassner, 1979.

Groffen, Mayke, *Lekker fris!: Honderd jaar gezondheid, schoonheid en fatsoen*, Amsterdam: SUN, 2009.

Grooss, K.S. and Tim Huisman, *Van piskijkers en heelmeesters: Genezen in de Gouden Eeuw*, Leiden: Museum Boerhaave, 2000.

Hamilton, James, *Observations on the Utility and Administration of Purgative Medicines in Several Diseases*, Edinburgh: Archibald Constable & Co and James Simpson/London: John Murray /Dublin: Gilbert and Hodges, 1811.

Hansard, Peter and Burton Silver, *What Bird Did That?: The Comprehensive Field Guide to the Ornithological Dejecta of Great Britain and Europe*, London: Grub Street, 1991.

Harington, John, *The Metamorphosis of Ajax: A Cloacinean Satire*, New York: Columbia University Press, 1962.

Hart, Maarten 't, *Het dovemansorendieet: Over zin en onzin van gewichtsverlies*, Amsterdam/Antwerp: De Arbeiderspers, 2007.

Haslam, Nick, *Psychology in the Bathroom*, Basingstoke/New York: Palgrave Macmillan, 2012.

Hauser, Albert, *Das Sitzpinkel-Manifest: Hier sitzt MANN*, Frankfurt am Main: Eichborn Verlag, 1997.

Hellema, Henk, *Geur en gedrag*, Amsterdam: De Brink, 1994.

Hennig, Jean-Luc, *Brève histoire des fesses*, Cadeilhan: Zulma, 1995.

Herz, Rachel, *That's Disgusting: Unraveling the Mysteries of Repulsion*, New York/

London: W.W. Norton & Company, 2012.

Holmes, Caroline, *The Not So Little Book of Dung*, Stroud: Sutton Publishing, 2006.

Holzwarth, Werner and Wolf Erlbruch, *The Story of the Little Mole who knew it was none of his business*, London: Pavilion Books, 2001.

Horan, Julie, *Sitting Pretty: An Uninhibited History of the Toilet*, London: Robson Books, 1998.

Houttuyn, Martinus, *Natuurlyke historie of uitvoerige beschryving der dieren, planten en mineraalen, volgens het samenstel van den heer Linnaeus*, Amsterdam: F. Houttuyn, 1761.

Hugo, Victor, *Les misérables*, Dutch translation by Manuel Serdav, Utrecht: A.W. Bruna, 1990.

Hunold, Günther, *Abarten des Sexualverhaltens: Ungewöhnliche Erscheinungsformen des Trieblebens*, Munich: Wilhelm Heyne, 1978.

Inglis, David, *A Social History of Excretory Experience: Defecatory Manners and Toiletry Technologies*, Lewiston, NY/Queenston, Ontario/Lampeter, Wales: Edwin Mellen Press, 2000.

Jansen, Gemma C.M., Ann Olga Koloski-Ostrow and Eric M. Moormann (eds.), *Roman Toilets: Their Archaeology and Cultural History*, Leuven/Paris/Walpole, MA: Peeters, 2011.

Jenkins, Joseph, *The Humanure Handbook: A Guide to Composting Human Manure*, White River Junction, VT: Chelsea Green, 2005.

Jong, Ellen, *Pees on Earth*, New York: PowerHouse Books, 2006.

Josephus, Flavius, *The Whole Genuine Works of Flavius Josephus*, translated by William Whiston, Glasgow: Blackie, Fullarton, and Co., 1829.

Kabakov, I., *The Toilet*, Kassel: Dokumenta IX, 1992.

Karlen, Arno, *Biography of a Germ*, New York: Pantheon Books, 2000.

Kavet, Herbert I. and Martin Riskin, *The Unspeakable Fart Book*, Watertown, MA: Boston American Corp., 1977.

Kemp, Christopher, *Floating Gold: A Natural (And Unnatural) History of Ambergris*, Chicago/London: University of Chicago Press, 2012.

King, Stephen, *On Writing: A Memoir of the Craft*, New York: Scribner, 2000.

Kira, Alexander, *The Bathroom: Criteria for Design*, New York: The Viking Press, 1976.

Kittelmann, Udo (ed.), *Ca-ca poo-poo*, Ostfildern-RFrom: Hatje Cantz, 1997.

Komrij, Gerrit, *Papieren tijgers*, Amsterdam: De Arbeiderspers, 1978.

Komrij, Gerrit, *Komrij's Kakafonie: Oftewel encyclopedie van de stront*, Amsterdam: De Bezige Bij, 2006.

Koster, S., *Bedwateren en verwante toestanden*, Haarlem: De erven F. Bohn, 1950.

Kotis, Greg and Mark Hollmann, *Urinetown: The Musical*, New York: Faber and Faber, 2003.

Kramer, Reinhold, *Scatology and Civility in the English-Canadian Novel*, Toronto/Buffalo/London: University of Toronto Press, 1997.

Krepp, Frederick Charles, *The Sewage Question: Being a General Review of All Systems and Methods Hitherto Employed in Various Countries for Draining Cities and Utilising Sewage*, London: Longmans, Green, and Co, 1867.

Kromhout, Rindert and Eric Smaling, *De prijs van poep*, Amsterdam: Leopold, 2001.

Kundera, Milan, *The Unbearable Lightness of Being*, translated by Michael Henry Heim, New York: Harper & Row, Publishers, Inc., 1984.

Lamarcq, Danny, *Het latrinaire gebeuren: De geschiedenis van de w.c.*, Ghent: Stichting Mens en Kultuur, 1993.

Lambton, Lucina, *Temples of Convenience & Chambers of Delight*, Stroud: Tempus Publishing, 2007.

Laporte, Dominique, *Histoire de la merde: Prologue*, Paris: Christian Bourgois, 1993.

Lee, Jae Num, *Swift and Scatological Satire*, Albuquerque: University of New Mexico Press, 1971.

Lerous, Pierre, *Aux Etats de Jersey*, Paris, 1853, quoted in Dan Sabbath and Mandel Hall, *End Product: The First Taboo*, New York: Urizen Books, 1977.

Lewin, Ralph, *Merde: Excursions in Scientific, Cultural, and Sociohistorical Coprology*, Aurum Press, London, 1999.

Linden, David J., *Genot als kompas: Hoe ons brein maakt dat we vet voedsel, orgasmes, sporten, marihuana, wodka, leren en gokken zo prettig vinden*, Amsterdam: Uitgeverij Nieuwezijds, 2011.

Ljung, Magnus, *Swearing: A Cross-Cultural Linguistic Study*, Houndmills/New York: Palgrave Macmillan, 2011.

Llosa, Mario Vargas, *In Praise of the Stepmother*, translated by Helen Lane, London: Faber and Faber, 1991.

Magdelijns, J.R.M. et al. (eds.), *Het staat op straat*: Straatmeubilair in Nederland,

The Hague: VNG-Uitgeverij/Sdu, 1990.

Mattelaer, Johan, *Seksuologische en minder logische verhalen: Liefde, lust en lichaam*, Leuven: Davidsfonds Uitgeverij, 2013.

McLaughlin, Terence, *Coprophilia: or, A Peck of Dirt*, London: Cassell, 1971.

Melville, Herman, *Moby-Dick*, New York: Harper & Brothers, 1851.

Merle, Ditte and Philip Hopman, *Lekker vies!: Alles over drupneuzen, stinktenen, smeeroren, zweet- handen en meer van die dingen*, Houten: Van Reemst, 1998.

Metchnikoff, Elie, *The Nature of Man: Studies in Optimistic Philosophy*, Honolulu: University Press of the Pacific, 2003/1908.

Miller, Susan B., *Disgust: The Gatekeeper Emotion*, Hillsdale, NJ/London: The Analytic Press, 2004.

Miller, William Ian, *The Anatomy of Disgust*, Cambridge, MA/London: Harvard University Press, 1997.

Moalem, Sharon, *Het nut van ziekte*, Amsterdam: De Bezige Bij, 2007.

Morris, Desmond, *Baby's*, Van Holkema & Warendorf: Houten, 1992.

Mozart, Wolfgang Amadeus, *Mozart's Letter, Mozart's Life*, edited and translated by Robert Spaethling, New York: W.W. Norton & Company, 2005.

Mullan, Desmond, *Farting: Gas Past, Present & Future*, Chatsworth, CA: CCC Publications, 1998.

Muth, Robert, *Träger der Lebenskraft: Ausscheidungen des Organismus im Volksglauben der Antike*, Vienna: Robert M. Rohrer, 1954.

Nesse, Randolph M. and George C. Williams, *Why We Get Sick*, New York/Toronto: Times Books, 1994.

Nohain, Jean and F. Caradec, *Le Petomane 1857–1945: A Tribute to the Unique Act Which Shook and Shattered the Moulin Rouge*, London: Souvenir Press, 1967.

Nussbaum, Martha C., *Hiding from Humanity: Disgust, Shame, and the Law*, Princeton/Oxford: Princeton University Press, 2004.

Obrist, Hans-Ulrich (ed.), *Cloaca Maxima*, Ostfildern bei Stuttgart: Cantz, 1994.

Ochs, Thomas, *Cornelius Kolig, Ein 'Paradies': Medientheoretischer Kontext*, Munich: GRIN, 2010.

Oort, Gert van, *De vos*, Utrecht/Antwerp: Het Spectrum, 1978.

Orwell, George, *The Road to Wigan Pier*, 1937, cited in William Ian Miller, *The Anatomy of Disgust*, Cambridge, MA: Harvard University Press, 1997.

Pagett, Matt, *What Shat That?: The Pocket Guide to Poop Identity*, Berkeley, CA: Ten Speed Press, 2007.

Paullini, Christian Frantz, *Neu-vermehrte, heylsame Dreck-Apotheke: Wie nemlich mit Koth und Urin fast alle, ja auch die schwerste, gifftigste Kranckheiten, und bezauberte Schäden vom Haupt bisz zum Füssen, inn- und äusserlich, glücklich curiret worden*, Frankfurt am Main: Friederich Daniel Knochen, 1696 (Facsimile Konrad Kölbl, Munich, 1969).

Pavlov, Ivan P., *De fysiologie van de spijsvertering: Acht colleges; een experimentele benadering*, Amsterdam: Candide, 1997 (original German edition 1898).

Persels, Jeff and Russell Granim (eds.), *Fecal Matters in Early Modern Literature and Art: Studies in Scatology*, Aldershot/Burlington, VT: Ashgate, 2004.

Philippe, Adrien, *Histoire des apothicaires chez les principaux peuples du monde*, Paris, 1853.

Pieper, Werner (ed.), *Das Scheiss-Buch: Entstehung, Nutzung, Entsorgung menschlicher Fäkalien*, Löhrbach: Der grüne Zweig 123, 1987.

Piras, Susanne (ed.), *Latrines: Antieke toiletten, modern onderzoek*, Meppel: Edu'Actief, 1994.

Pleij, Herman, *Het gilde van de Blauwe SchFrom: Literatuur, volksfeest en burgermoraal in de late middeleeuwwen*, Amsterdam: Meulenhoff, 1979.

Pleij, Herman, *Dromen van Cocagne: Middeleeuwse fantasieën over het volmaakte leven*, Amsterdam: Prometheus, 1997.

Pokuss, Gil, *Geliebtes Schwein*, Frankfurt am Main: Melzer Verlag, s.a.

Praeger, Dave, *Poop Culture: How America is Shaped by its Grossest National Product*, Los Angeles: Feral House, 2007.

Promblés, E. Slove, *Fee! fie! foe! fum!: A Dictionary of Fartology*, Rockport, NY: Melodious Publications, 1981.

Promblés, E. Slove, *Flatulence Denied!: Stalking the Fartological Imperative*, Rockport, NY: Melodious Publications, 1983.

Provine, Robert, R., *Curious Behavior: Yawning, Laughing, Hiccupping, and Beyond*, Cambridge, MA/London: Belknap Press, 2012.

Putman, Roderick, *Carrion and Dung: The Decomposition of Animal Wastes*, London: Edward Arnold, 1983.

Queri, Georg, *Kraftbayerisch: Wörterbuch der erotischen und skatologischen Redensarten der Altbayern*, Munich: Deutscher Taschenbuch Verlag, 1981 (original edition 1912).

Rabelais, François, *Gargantua and Pantagruel, Complete*, in *The Works of Rabelais*, Faithfully Translated from the French by Sir Thomas Urquhart of Cromarty and Peter Antony Motteux, Derby, The Moray Press, 1894. (Project Gutenberg).

Rabkin, Eric S. and Eugene M. Silverman, *It's a Gas: A Study of Flatulence*, Riverside, CA: Xenos Books, 1991.

Rauch, Erich, *De Mayr-darmreinigingskuur... en daarna gezonder leven: Darmreiniging, ontslakking, gezondere voeding*, Amsterdam: De Driehoek, 1992.

Reid, Donald, *Paris Sewers and Sewermen: Realities and Representations*, Cambridge, MA/London: Harvard University Press, 1991.

Reyburn, Wallace, *Flushed with Pride: The Story of Thomas Crapper*, London: MacDonald and Company, 1969.

Reynolds, Dan, *The Toilet Zone: A Hilarious Collection of Bathroom Humor*, Chatsworth, CA: CCC Publications, 1999.

Reynolds, Reginald, *Cleanliness and Godliness: or, The Further Metamorphosis*, New York: Doubleday and Company, 1946.

Richman, Josh and Anish Sheth, *What's Your Poo Telling You?* San Francisco, CA: Chronicle Books, 2007.

Roche, Charlotte, *Wetlands*, translated by Tim Mohr, London: Fourth Estate, 2009.

Roebroeks, Wil (ed.), *Guts and Brains: An Integrative Approach to the Hominin Record*, Leiden: Leiden University Press, 2007.

Roessiger, Susanne and Heidrun Merk (eds.), *Hauptsache gesund!: Gesundheitsaufklärung zwischen Disziplinierung und Emanzipation*, Marburg: Jonas Verlag, 1998.

Sabbath, Dan and Mandel Hall, *End Product: The First Taboo*, New York: Urizen Books, 1977.

Sale, Charles, *The Specialist*, London: Putnam & Company, 1930.

Schramm, Heinz-Eugen, *L.m.i.A.: Des Ritters Götz von Berlichingen denkwürdige Fensterrede*, Tübingen am Neckar: Fritz Schlichtenmayer, 1960.

Schuilenga, J.H. et al. (eds.), *Honderd jaar telefoon: Geschiedenis van de openbare telefonie in Nederland 1881-1981*, The Hague: PTT Telecommunicatie, 1981.

Scott, James, *Commentaries on the Use and Necessity of Lavements in the Correction of Habitual Constipation, and in the Treatment of Those Diseases Which are Occasioned or Aggravated by Intestinal Accumulation and Irritation*, London: J. Churchill, 1829.

Scott, Munroe, *Oh, Vulgar Wind: A Sympathetic Overview of the Common Fart*, Toronto: Culture Concepts, 1994.

Seltzer, Barry and Erwin Seltzer, *Farting: Ripper, Toot, Vapor, Passing Gas... A Fart by Any Other Name Would Not Smell as Sweet!* Scarborough, Ontario: Prism Publishing, 1999.

Shrager, Sidney, *Scatology in Modern Drama*, New York: Irvington Publishers, 1982.

Simmons, Joe J., *Those Vulgar Tubes: External Sanitary Accommodations Aboard European Ships of the Fifteenth Through Seventeenth Centuries*, College Station, TX: Texas A&M University Press, 1997.

Slijper, E.J., *Walvissen*, Amsterdam: D.B. Centen, 1958.

Smit, Raphaël, *Gemakshalve: 150 jaar riolering in Amsterdam*, Amsterdam: Dienst Waterbeheer en Riolering, 2000.

Smout, Andreas J.P.M., *Darmgewoel en darmgevoel*, Utrecht: Universiteit van Utrecht, 1994.

Smout, A.J.P.M. and L.M.A. Akkermans, *De normale en gestoorde bewegingen van het maagdarmkanaal*, Tilburg: Janssen Pharmaceutica, 1991.

Smout, A.J.P.M., *Prikkelbare darm syndroom*, Wormer: Inmerc, 1995.

Snijdewind, A. (ed.), *Als poepen niet gewoon gaat: Ontlastingsincontinentie (encopresis)*, Badhoevedorp: Mension Medical Refresher, 2005.

Sparnaay, M.J., *Van spierkracht tot warmtedood: Een geschiedenis van de energie*, Den Bosch: Voltaire, 2002.

Spinrad, Paul, *The RE/Search Guide to Bodily Fluids*, San Francisco, CA: re/Search Publications, 1994.

Stoddard, D. Michael, *The Scented Ape: The Biology and Culture of Human Odour*, Cambridge, etc.: Cambridge University Press, 1990.

Süskind, Patrick, *Perfume: The Story of a Murderer*, translated by John E. Woods, New York: Alfred A. Knopf, 1986.

Swaab, Dick, *We Are Our Brain: From the Womb to Alzheimer's*, translated by Jane Hedley-Prôle, London: Penguin, 2015.

Swift, Jonathan, *Gulliver's Travels*, London: Benjamin Motte, 1726.

Tanizaki, Jun'ichiro, *In Praise of Shadows*, translated by Thomas J. Harper and Edward G. Seidensticker, Stony Creek, CT: Leete's Island Books, 1977.

Teigetje and Woelrat, *Ons leven met Reve*, Amsterdam: Balans, 2012.

The Gastric Regions, and Victualling Department: By an Old Militia Surgeon, London: Robert Hardwicke, 1861.

Thomas, Carmen, *Ein ganz besonderer Saft: Urin*, Cologne: Vgs, 1993.

Thomas, Lewis, *The Lives of a Cell: Notes of a Biology Watcher*, New York: The Viking Press, 1974.

Topor, Roland, *Vinci avait raison*, Paris: Christian Bourgois Editeur, 1976.

Toscani, Oliviero, *Cacas: Die Enzyklopädie der Kacke/The encyclopedia of poo/L'encylopédie*, Cologne: Taschen, 1988.

Turner, J. Scott, *The Extended Organism: The Physiology of Animal-Built Structures*, Cambridge, MA/London: Harvard University Press, 2000.

Vetten, Horst, *Über das Klo: Ein Thema, auf das jeder täglich kommt*, Luzern/Frankfurt am Main: C.J. Bucher, 1980.

Vigarello, Georges, *Le propre et le sale: L'hygiène du corps depuis le Moyen Âge*, Editions du Seuil, 1985.

Vogel, Klaus (ed.), *Das Deutsche Hygiene-Museum Dresden 1911–1990*, Dresden: Michel Sandstein Verlag, 2003.

Vogel, Steven, *Vital Circuits: On Pumps, Pipes, and the Workings of Circulatory Systems*, New York, etc.: Oxford University Press, 1992.

Voûte, A.M. and C. Smeenk, *Vleermuizen*, Zwolle: Waanders Uitgevers, 1991.

Vries, Leonard de, *Ha dokter ho dokter: knotsgekke geneeskunde uit grootvaders tijd*, Bussum: De Haan, 1974.

Vroon, Piet, Anton van Amerongen and Hans de Vries, *Verborgen verleider: Psychologie van de reuk*, Baarn: Ambo, 1994.

Waal, M. de, *Dieren in de volksgeneeskunst*, Antwerp: De Vlijt, s.a.

Waltner-Toews, David, *The Origin of Feces: What Excrement Tells Us About Evolution, Ecology, and a Sustainable Society*, Ontario: ECW Press, 2013.

Wateetons, Meneer and Sjoerd Mulder, *Over worst*, Amsterdam: Carrera, 2011.

Watson, Lyall, *Jacobson's Organ and the Remarkable Nature of Smell*, New York/London: W.W. Norton & Company, 2000.

Werner, Florian, *Die Kuh: Leben, Werk und Wirkung*, Munich: Nagel & Kirche, 2009.

Wetzel, Donald, *The Complete Joel's Journal and Fact-Filled Fart Book*, Bayside, NY: Planet Books, 1983.

Wheeler, Billy Edd, *Outhouse Humor*, Little Rock, AR: August House, 1988.

Whitfield, John, *In the Beat of a Heart: Life, Energy, and the Unity of Nature*, Washington, DC: Joseph Henry Press, 2006.

Whorton, James C., *Inner Hygiene: Constipation and the Pursuit of Health in Modern Society*, Oxford, etc.: Oxford University Press, 2000.

Wolfe, David W., *Tales from the Underground: A Natural History of Subterranean Life*, Cambridge, MA: Perseus Publishing, 2001.

Woud, Auke van der, *Koninkrijk vol sloppen: Achterbuurten en vuil in de negentiende eeuw*, Amsterdam: Bert Bakker, 2010.

Wouterlood, Floris, *Stof zijt gij*, Uithoorn: Karakter Uitgevers, 2003.

Wright, Lawrence, *Clean and Decent: The Fascinating History of the Bathroom & the Water Closet*, London: Routledge & Kegan Paul, 1960.

Wright, R.H., *The Science of Smell*, London: George Allen & Unwin, 1964.

Wynn, Makin, *Of Pots and Privies*, Middelburg, VA: Derlinger's, 1959.

Zglinicki, Friedrich von, *Geschichte des Klistiers: Das Klistier in der Geschichte der Medizin, Kunst und Literatur*, Frankfurt am Main: Viola-Press, s.a.

Zimmer, Carl, *Parasite Rex: Inside the Bizarre World of Nature's Most Dangerous Creatures*, New York, etc.: The Free Press, 2000.

Zimmer, Carl, *Microcosm: E. coli and the New Science of Life*, New York: Pantheon Books, 2008.

Zinsser, Hans, *Rats, Lice and History*, New York: Little, Brown and Company, 1935.

Illustration Credits

p.3   Nederlands Tegelmuseum, Otterlo, 1625–1650.

p.6   Still from *Le Fantôme de la liberté*, Luis Buñuel, 1974.

p.13   Photononstop RM/Hollandse Hoogte.

p.35   Engraving from KaPiFu, 1556.

p.37   From: Reinier de Graaf. De clysteribus. Leiden, 1668.

p.49   Engraving from KaPiFu, 1545.

p.58   Bettmann/Corbis.

p.76   Carthage, Tunisia. Michael Nicholson/Corbis.

p.86   Leonardo da Vinci. From: *Atlas der anatomischen Studien*. AKG/ANP.

p.89   Still from *Kreatief met Kurk*, VPRO.

p.91   French caricature. Science Photo Library/ANP, ca. 1770.

p.118   Hieronymus Bosch. *Garden of Earthly* Delights. Museo del Prado.

p.123   Getty Images.

p.145   German army field latrine regulations, April 1916. From: Englisch, 1928.

p.154   Woodcut from J. de Damhoudere: *Practycke ende handtbouck in criminele zaeken*, Leuven, 1555. From: Lamarcq, 1993.

p.161   Spaarnestad Photo.

p.162   AKG/ANP.

p.176   Mediaeval woodcut. From: Pieper, 1987.

p.183   Rajasthan, India. AKG/ANP.

p.190   Advertising material, kopi luwak.

p.225   Watercolour by Jan Augustini, Philips Wouwerman. Rijksmuseum Amsterdam, 1759.

p.230   Anti-farting sign, AKG/ANP.

p.237   Engraving by F. de Bakker in J.A. de Chalmot: *Algemeen woordenboek*, 1778. From: Zglinicki, s.a.

p.257   Print by Thomas Rowlandson. Mary Evans Picture Library, ca. 1800.

p.263   From: Ebberfeld, 1998.

p.265   From: *Autre dissertation sur le même sujet* (1743). From: Komrij, 2006.

Index

A

Adam and Eve 53, 141

alcohol 12, 94, 119, 220

Allais, Alphonse 226

Allport, Gordon 26

ambergris 165–9, 189

amino acid 119

ammonia 171, 178, 184, 222–4

andouillette 120

antibiotics 99, 100, 130

antiquity 70, 76, 162, 186, 230

anus 11, 25, 34, 41, 58, 72, 89, 91, 93, 95, 96, 108, 122, 128, 130, 133, 146, 148, 193–216, 230–1, 233, 236, 238, 244, 246–8, 266

appendix 14, 88, 98, 107, 122, 196, 248

Arbuthnot, Lane 108

Aristophanes 155, 241

Aristotle 181

arse 7, 41, 44, 48, 57, 68–9, 125, 135, 145, 148, 213, 231, 241–4, 247–8, 253

Ashton, T.J. 147

astronaut 58, 66

B

baby 5, 10, 19, 21–2, 28, 69, 70, 72–3, 100, 105, 117, 138, 153, 205, 229, 252–3, 262

back stairs, French 58

Bacon, Sir Francis 254

bacteria 46–7, 71–2, 78, 92, 99–101, 106–8, 111, 115, 119, 123, 126, 128–30, 152, 159, 176, 178–9, 182, 184, 224–5, 231, 238–41, 263–4

Balzac, Honoré de 8

Bauhin, Caspar 34

Beatrix, Queen 187

Beaumont, William 93

bedpan 20, 32

bicarbonate 95–6, 231

bile 73, 87–8, 96, 102–3, 115, 225

bilirubin 115, 269

birth 10, 70, 73, 100, 136–9, 143–4, 149, 251, 253

bladder 36, 70, 72, 75, 102, 138, 142, 147, 153, 199, 218, 222–3, 226–7, 257

Blaxter, Kenneth 84

blood 13, 17, 19, 32, 35–6, 40, 53, 59, 63, 72–3, 84, 90, 94, 96, 101, 108, 111, 114–5, 118, 130, 142, 148, 152, 173, 220–1, 225, 229, 231, 269

Blum, Gary 117

Böll, Heinrich 262

Bomans, Godfried 215

borborygmus 229

Bosch, Hieronymus 118

bottom 7, 11, 13, 21, 31, 37–8, 43–4, 46, 58–9, 64, 146, 148, 154, 176, 195, 220, 247, 252, 258

Bouchard, Charles 152
brain 13–18, 28, 32, 43, 52, 80–1, 99, 103–4, 106, 114, 136, 142–3, 146, 149, 165, 175, 197, 205, 207, 210, 214, 220, 226, 253
Brezhnev, Leonid 261
Brooks, Mel 239
Brown, Norman 254
Buckland, Dean 134
Buñuel, Luis 5, 6
Bush, George W. 261

C

caca de dauphin 117
call of nature 11, 140
carminative 240
carnivore 105–6, 108, 118
Casa, Giovanni della 52
casing 119–22, 190
catharsis 12–13, 137, 226
cesspool 160, 162, 192 269
chamber pot 24, 57, 154, 214–5, 257, 268
Chappell, George S. 87–9
Chaucer, Geoffrey 242
Chesterfield, Lord 42
childbirth 10, 19, 137–9
cholera 46, 160, 176
cistern 58–9, 224
Cloaca Maxima 162
Clostridium difficile 100
clyster 34, 151

colon 13, 39, 108
Comfort, Alex 206
Commodus, emperor 264–5
compost toilet 176
constipation 49, 144, 149–52, 167
coprophagia 265–6
coprophile 134, 254
corpse 71–3, 87
Courbet, Gustave 267
Crapper, Thomas 59
Crichton, A. 134
cud, chewing 107–8

D

Da Vinci, Leonardo 86, 105
Darwin, Charles 26–7, 32, 153, 180–2
De Goncourt, Edmond and Jules 164
De Graaf, Reinier 36–7
De Sade, Marquis 265
deep flusher 59
defecatory posture 143
Delvoye, Wim 112
Desai, Moraji 224
diarrhoea 20, 39, 121, 123, 125–6, 130, 238, 265
digestion 9, 26, 86, 92–5, 101–2, 112, 142, 156, 261
Diogenes, philosopher 267
Douglas, Mary 25, 50
drainage 58, 144, 163–4
drugs 225, 240

Dubois, Eugène 172

duodenum 95–6, 102

dung beetle 155–9, 188

dung flies 29, 157–9

E

earth closet 59

earthworm 130, 159, 181–2, 191

eating 3–7, 10, 14, 16, 18, 20, 26–8, 30, 32, 36, 39, 48, 54, 77, 81, 84, 102, 106, 109, 114, 117–8, 125, 127, 129, 131–2, 135, 141, 149, 151–2, 156, 158–9, 174, 177, 182–3, 189, 219, 229, 238, 240–1, 261, 264, 266–7

Egyptians, ancient 17, 34, 156, 188

Eibl-Eibesfeldt, Irenäus 27

Einstein, Albert 16

Ekman, Paul 27

Elias, Norbert 52

Elizabeth I, Queen 58

endoscope 90, 92

enema 34–9, 150, 152

energy 10, 15, 17, 81–5, 96, 99, 104–8, 174, 177, 179, 183–4, 189

enzyme 72, 86, 95–6, 99, 107, 190, 238, 240

Erasmus, Desiderius 52–3

Escherichia coli 129

evolution 22, 26, 65, 70, 83, 97–9, 107, 144, 148, 181, 196, 228

Eza'e, Kirupano 22

Ezekiel, prophet 266

Fabre, Jean-Henri 155–6

F

faecal transplantation 130

faecal/faeces particle 29, 47

faeces
 as art 110–11
 as a means of communication 119
 as medicine 100, 130, 185–6
 as souvenir 191

fart 53, 84, 99, 112, 125, 179, 198, 213, 230–9, 241–4, 247, 256, 259–60, 262, 269

Feable, Josiah 76

fermentation 106–8, 231

fibre 36, 49, 57, 122–3, 125, 127, 129, 132, 135, 149, 224, 226, 231, 238–9

First World War 142, 145, 178–9, 245

flatulence 235, 239–41

Français, François-Louis 267

Freud, Sigmund 12, 251–5, 257

fullones (fullers) 223

G

Gagnaire, Pierre 2

Galen, physician 186

gall bladder 15

Gandhi, Mahatma 224

Garbo, Greta 105

Gargantua and Pantagruel 43

gas 5, 29–30, 119, 146, 153, 160–1, 178, 184, 216–45, 256, 259

gastric juice 92–3, 95, 102, 231
gastroenterologist 90
Gayetty, Joseph 42
Gershon, Michael 103–4
God 18, 24, 34, 47–9, 62, 83, 122, 137, 139, 141, 146, 155, 187, 212–3, 229–30, 243–4, 251, 266–7
Goethe, Johann Wolfgang von 213
Goldberg, Whoopi 236
Gorbachev, Mikhail 261
Graves, Dr W.H. 152
Greeks, ancient 84, 224, 238, 241
greenhouse effect 184, 235–6
guano 170–2, 178
gut 3, 9, 15, 43, 81, 103, 106, 122, 129–31, 147

H

Haber-Bosch process 179
haemoglobin 115, 118
haemorrhoid 42, 144, 148, 233
haemorrhoidal plexus 148
Hagen, Albert 250
Hall, Mandel 141, 143
Halting Stations 76
Hansard, Peter 133
Harington, Sir John 58
Hartogh Heys van Zouteveen, Hermanus 32
Haussmann, George 163
Heine, Heinrich 186, 268

Hendrik, prince of the Netherlands 161
Henry III, King 64
Heracles 163
herbivore 105–8, 132, 135
Herodotus, historian 234
hierarchy 14–5, 94
Hitler, Adolf 50, 179
Hofland, Henk 25
hormones 63, 82, 140, 142–3, 205, 210–1
Houttuyn, Martinus 185
Huet, G.D.L. 51
Hugo, Victor 173
human manure 174–5, 179, 184, 192
Humboldt, Alexander von 170
hunger 12, 14, 32, 94, 104, 106, 125, 178, 229
Hunold, Günther 254, 264
Huntington's disease 28
hydrochloric acid 93–4
hydrogen 119, 178, 222, 231–2
hygiene 54, 129, 160

I

illness 38–9, 90, 129, 152–3, 160, 224, see also: sickness, disease, illness
immune system 29, 72, 99, 132
incontinence 72, 74, 144, 153, 227
insula 28, 264
intestinal bacteria 99–111, 119, 129, 241
intestinal flora 99, 100, 119, 129–30

intestinal tract 73, 91, 95, 103, 127, 131, 190, 220, 229

intestinal wall 15, 96, 98, 122, 149, 231–2

intestine 17, 34, 45, 48, 73, 85–7, 90, 95–101, 104–8, 115, 118–20, 122–3, 126–9, 131, 139–140, 144, 146–9, 152–3, 167, 184, 220, 230–1, 238, 240, 248, 261, 263

large 101, 106–8, 120, 123, 126, 139–40, 149, 152, 184, 220, 231, 238, 240, 248

length 85, 96, 98

small 95–6, 98–101, 104, 106–7, 115, 120, 122, 131, 238, 248

intestines 4–5, 8–9, 12, 14–18, 20, 24, 34–6, 39–40, 45, 48, 57, 60, 65–6, 72–3, 81, 83, 86–90, 90, 94–108, 111–15, 118, 120–22, 126–7, 129, 131, 140–42, 144, 149–53, 167, 181, 190, 220, 229, 236, 238, 251, 263–4

J

James, William 50

Jansen, Gemma 77

Jenkins, Joseph 177, 192

Jones, Andrew 135

Josephus, Flavius 234

Joyce, James 249

K

Kellogg, Dr John Harvey 108, 140, 152, 208

Kemp, Christopher 168

kidneys 8, 17, 115, 127, 142, 199, 218, 220–2, 226

King, Stephen 256

Klenk, Hans 42

koala manure 191

Koch, Robert 45, 160

krul 77–8

Krünitz, Johann Georg 35, 38

Kundera, Milan 48

L

laxative 90, 116, 149–50, 152–3, 265

Leahy, Michael 22

Lennon, John 224

Leo V, Pope 64

Leroux, Pierre 174

Lewin, Ralph 133, 142, 175

Liebig, Justus von 177

Liernur System 161

Liernur, Charles 160

Linnaeus, Carl 210

litter box 20, 53, 207, 253

liver 72, 87, 96, 114, 115, 130, 212

Louis XIV 37, 42

Louis, Rudolf 42

Luther, Martin 48–9, 119, 243

M

Maintenon, Madame de 42

Manzoni, Piero 110

Marie-Antoinette 117

Marshall, Victor 146
meconium 73
Melville, Herman 166
membrane 29, 108
methane 184, 231–2, 235–6
microbe 45–6, 87, 99, 152, 176, 236
Middle Ages 42, 52, 224, 247
Miller, William 30–1, 41
molecule 80, 96, 115–6, 203–4
Molière, Jean Baptiste 38
Montaigne, Michel de 230
Morrison, Jim 224
mould 129, 181, 256, 265
Moule, Henry 59
Mozart, Wolfgang Amadeus 70, 243, 247
Mulisch, Harry 187–8
mummify 17, 170
Murie, Olaus 134
myoglobin 115–5, 118

N

Napoleon 70, 263
nappy 5, 19, 11
nematode 159, 182
neon tetra 20
nervous system 9–13, 102–4, 140, 175, 226–7
   autonomic 9–11, 102–3, 226–7
   enteric 103
   parasympathetic 102, 103
   sympathetic 102, 156

nitrogen 116, 168–9, 178–9, 181–2, 222, 231–2
nutrient 97, 104, 106–8, 126–7, 174, 179, 189, 264

O

Ockels, Wubbo 66
oesophagus 73, 87, 91, 95, 229
oligosaccharides 238–40
omnivore 107–8
orifice 3, 11, 32, 34–5, 121, 251
Orwell, George 51
oxygen 65, 86, 88, 99, 115–6, 184, 222, 224, 228, 231–2
oxymyoglobin 115

P

paramecium 136
parasite 131–2, 142–3, 159–60
perfume 165–6, 168, 186, 198, 201–4, 206
Parker, G.A. 159
Pasteur, Louis 160
Paullini, Christian 185
Pavlov reflex 227
pee diet 127
peeing in public 227
peristalsis 101, 104
Petomane, le (Joseph Pujol) 244–5
Peyerl, Johann 247
Philippe, Adrien 36
phosphorus 168–9, 181, 190–1

pinworm 130
plankton 170, 189
plateau toilet 59
poo clinic 154
Porphyry, philosopher 141
Praeger, Dave 49
Prikkeprak the Gnome 88–9
privacy 56, 59, 68, 249
probiotics 99
Protagoras's statement 62
protein 95–6, 113, 119, 121, 127, 142, 175, 190, 222, 225, 238–9
puke 54, 90–2, 95, also see: vomit

R

Rabelais, François 43, 150
Ranby, Mr John 146
rectum 34, 70, 88–9, 120, 123, 128, 130, 138–40, 144–7, 152, 236, 248, 250
Reeves, Jim 74
Reger, Max 42
Reve, Gerard 6, 7, 259
Richelieu, Cardinal 43
Roche, Charlotte 53, 248, 250
Romans 44, 92, 162–3, 223, 264
Roth, Philip 249
roundworm 131
Rousseau, Jean Jacques 202
Rozin, Paul 32
rubbish 11, 47, 56, 65, 68, 112, 138, 160, 173, 176, 184, 217, 267

S

Sabbath, Dan 141, 143
salt 24, 39, 98–9, 111, 127, 168, 177–9, 204, 219–21
Sanisette 258
sanitary facilities 72, 77
sausage 113–16, 118–23, 128–9, 262
Schäffer, Johann Gottlieb 35
Schmidt, Annie M.G. 196
senses 9–10, 12, 64, 104, 246, 249
sewer system 42, 54, 59
sex 1–3, 6–7, 13–15, 18–9, 54, 102, 135, 199, 201–2, 212, 214–6, 248–9
Shakespeare, William 50
shame 14, 36, 46, 51–3, 211, 223, 230, 239, 252, 255
shape 4, 7, 31, 59, 72, 127–8, 191, 246–7
Shaw, Bernard 105
shit diet 125, 127
skatole 119, 203–4, 232, 241
smell 204–5, 207–8, 210–11, 222, 224, 231–2, 239
Smyth, Andrew 171
Snow, Dr John 160
Socrates 241–2
spermaceti 167
sphincter 41, 91, 93, 95, 139, 144, 146, 154
squat toilet, French 57, 144
St. Martin, Alexis 93
stench 5, 19, 29–30, 54, 118–9, 126, 160,

164, 187, 198, 211, 232, 239
stercobilin 115
sterile 78, 100–1
stomach 9, 17–8, 25, 31, 40, 60, 65, 73, 86, 90–5, 97, 101–2, 104, 106, 127, 138, 140, 150, 167, 219, 229–30, 238, 241, 244
stool 150, 213, 261
sulphite 232
Süskind, Patrick 211
Swaab, Dick 15
sweating 9, 219–20
Swift, Jonathan 261

T

tapeworm 131
taste 5, 8–9, 11, 28–30, 32, 45, 122, 127, 129, 158, 170, 187, 190, 222, 224
taste bud 8–9, 122, 127, 246, 266
temperature 31, 120–1, 176
't Hart, Maarten 125–6
Thomas, Lewis 46
Thompson, D'Arcy 82
toilet 3–5, 10, 13–4, 19–20, 24–5, 28, 30, 33, 40–7, 54, 56–62, 64–72, 74–9, 91, 93, 111–3, 123–5, 128–9, 144–5, 149, 153–4, 161–2, 173, 176–7, 182, 186, 189, 192, 194, 201, 214, 216, 226–7, 230, 233, 249–50, 252–5, 257–9, 261, 264, 268–9
  Japanese 268
  public 44, 47, 76, 162
  Roman 76

toilet paper 5, 19, 42–4, 54, 250, 258
toilet seat 30, 59, 201
toilet-trained 154, 226–7, 252–4
touch 3, 21, 30, 62, 64–6, 76, 188, 203, also see: feel
Trygaeus 155
tupaya 196

U

urinal 77–8, 249
urine 224–8, 248–9, 251, 254, 264
urobilin 115
urophile 254

V

Van Dam, Johannes 2
Van der Geest, Sjaak 175
Van der Mey, Johan 78
Van Diepenbeek, Annemarie 134
Van Leeuwenhoek, Antonie 45
Van Maerlant, Jacob 34
Vargas, Llosa Mario 50
vegetarian 81, 105–6, 118, 123, 131, 241
Vespasian, Titus Flavius, Emperor 76, 223
vespasiennes 76
Vetten, Horst 60–1
Victoria, Queen 3, 76, 177
virus 46
vomit 25, 90–2, 95, 106, 166, also see: puke
vulture, American 196

W

waste 33, 72, 81, 94, 112, 114–5, 126–7, 138, 149, 160, 162–3, 173, 178, 180, 189, 197, 228

water 218–45, 258–9

weight 66, 81, 98, 124, 126–7, 131, 144, 168

Werner, Florian 186, 264

West 45, 144, 153, 174–5, 183–4

Y

Yeats, William 247

Z

Zeus 81, 241